MANAGING
THE
ENVIRONMENT

MANAGING
THE
ENVIRONMENT

AN ECONOMIC PRIMER

by

WILLIAM RAMSAY

and

CLAUDE ANDERSON

BASIC BOOKS, INC., PUBLISHERS

NEW YORK   LONDON

© 1972 by Basic Books, Inc.
Library of Congress Catalog Card Number: 72–76907
SBN–465–04377–1
Manufactured in the United States of America
DESIGNED BY THE INKWELL STUDIO

To all the members of our *karass:*
"Busy, busy, busy."

PREFACE

Ecology is an awfully good word to bandy about nowadays. It seems ages since the word meant the study of the interactions of the plants and animals in a peat bog. Instead, ecology seems to mean now just about anything that is wrong with the world. Before, it meant a neutral subject of study, something like history or geography, but now the word seems to bring forth the emotions usually attached to words such as "wisdom" and "honesty." And when the name of a subject of study becomes a catchword for all ills, it is natural that a lot of people are going to write a lot of books and articles telling off the enemies of the sacred cow involved, in this case, "ecology" or "environment." All this is fine if it advances the work of society. It is important that the public be aroused to deal with environmental problems, such as smog, water pollution, and all of the others. So we think that cataloging the sins of man against nature and exhorting people to do better does fulfill a useful purpose.

But we must confess that a creeping sense of tedium begins to overtake us in the middle of works explaining just how depressing the Florida Everglades look nowadays, or how all the helpless fish in the Great Lakes are doomed to imminent extinction. So we have neither the motivation nor the patience to write a "Why don't they do something?" book. Instead, we have tried to write a "Why don't they do such and such?" book, and, as far as we have been able to, we have tried to actually write a "How to . . . " book, in this case "How to use economic planning to mitigate the disruption of the environment."

So you will find the book rather short in catalogs of environmental troubles, although, of course, some do have to be mentioned for purposes of illustration. There is really no treatment here of ecology in the way an ecologist (biologist) treats the subject. Certainly, the subject matter of ecology in the pure sense is of public interest, because one of the big questions the world faces is how to make changes in the environment without destroying valuable ecological habitats. But since we are writing a book proposing practical measures, we necessarily give fairly limited coverage to questions, unless at least some answers to those questions, in terms of human actions, are known. We must emphasize that we do not think the most tractable problems are necessarily the most worthy. What does disturb us, and what gives us our economic bias in writing here, is that the possibilities for action on the tractable problems, such as air pollution, are not being fully realized. Some problems, such as the esthetic value of an untouched wilderness area, may actually deserve more attention than smog, for example. We think it inevitable that such difficult questions will receive more emphasis as time goes by. But time is of the essence, and let us not quarrel about which problem is most important. Action, not quibbling, is needed.

But even though it might be obvious that action is needed, why do we emphasize economics? Well, it seems to us that economics is what really counts, when people get down to essentials. Environmentalists must remember that "ecology" is like "virtue." When asked, everyone is for it; everyone says,

"Yes, yes, that's right." But when it comes down to daily life, then both sin and environmental disruption have their day. We know that the environmentalist, when he realizes this, often says, "Well, economics can go to the devil for all I care." From our admittedly biased point of view, we think that this result is neither possible nor desirable. What we do want to say to "anti-economists" is that when economics seems to give answers that are intuitively wrong, such as being on the side of the polluters of our rivers, then maybe economics isn't doing quite the job it should. This is not because economists have not considered these questions as generally used in theory. They have. But most economics as used by engineers and planners at the grass roots level does not adequately reflect the trends in theoretical economics. And, economics as a social science does naturally tend to find research concentrated in areas where there are more data, such as in the normal interchanges of the market economy. To treat pollution and other environmental problems correctly, one has to go outside the market economy as we understand it. This is the difficulty, since data for planning purposes are rather scanty, and theoretical methods of interpreting the data are somewhat uncertain. Despite these difficulties, we believe that the need for a rational, economic, planned methodological approach to the environmental disruption problem is so great we had better do something about it now, even with imperfect information and methods.

An economics primer for the environment cannot begin to give exact recipes for fixing up the world. But we have tried to describe the outlines of a definite planning scheme for government use, just to help you visualize what exactly it is we are recommending. Obviously, such schemes require much more study before anything practical is done. This last disclaimer applies not only to the government agencies suggested in the last chapter, but also to the sequence of planning steps explained in Chapter 9.

We have already indicated that there is no technical

treatment of questions of biology here. We have also tried to keep the discussion of economics in the main text as straightforward and nontechnical as we can. Chapter 4 serves as an orientation to the economic mode of thinking, while Chapter 5 emphasizes the peculiar nonmarket characteristics of pollution and other environmental problems.

For the serious student, the economic considerations of Chapters 4 and 5 and the planning decision models of Chapter 9 are illustrated or explained in the Appendices, in one case with the use of a modest amount of mathematics. This relegation of mathematics to the Appendices follows the American economist, George Stigler, who noted that mathematics in an appendix is much easier to ignore. Sufficient notes are included, we hope, so that students will be able to get a footing in the literature by consulting the references named. We hope that we can help to stimulate public and governmental interest in rational approaches to environmental problems. If the importance of a methodical approach to these problems can be impressed upon people, we think that the prospects for a liveable world in the future will be much brighter.

ACKNOWLEDGMENTS

Thanks to:

Robert Dorfman, Harvard University, for expert criticism of some economic and mathematical points.

William Blaine, for supplying motivation and encouragement and chapter-by-chapter commentary and suggestions.

J. B. deC. M. Saunders, University of California, whose encyclopedic knowledge saved us from making some gross errors and usefully challenged our logic in certain places.

Edward Friedland, State University of New York, Stony Brook, for his incisive comments and critical review of political implications of unequal distributions of income in ecological analysis.

Systems Associates, Inc., from one of the authors, for forbearance during the writing process, and its president, Lawrence Kavanau, for an introduction to new fields of study.

Pacific Architects and Engineers Incorporated and its forward-thinking management.

Sandra Bechtold, for her pugnacious attitude toward our syntax. We love her and the girls at the Posthouse anyway.

Constance Shainline, for all her help.

The women in our lives, Hildah and Margaret, for smoothing out with love the peaks and valleys of our ego cathection.

Of course, none of the above are to be held responsible for any factual errors or for controversial points of view.

CONTENTS

Contents

MANAGING
THE
ENVIRONMENT

1

THE TROUBLED
ECOLOGICAL CYCLE

It is not really a coincidence that the first two syllables of the two words, *ecology* and *economics,* are the same. Those two syllables come from the Greek word for "house" or "dwelling place," and the history of the two words gives an illustration of the growth of consciousness in man and of the extension of his intelligence and concern into wider and wider areas.

The older word, *economics,* first meant the science of management of a single household or farm and then was extended into the operations of businesses. Ultimately, it became the study of the production and exchange of goods in cities, nations, and the world at large. The newer word, *ecology,* was coined to describe the science of the exchange of energy and the interdependence of life among various types of animals and plants. From the basic consideration of the ecology or *ecosystem* (a more or less closed system of reciprocal life patterns) of grass-eating rabbits and rabbit-eating foxes, the concept has been subsequently enlarged to encompass the relations of populations of flora and fauna in sizable geographic areas, and most recently, to a consideration of the totality of

3

close interrelationships of life and matter in the world ecosystem.

We see in both cases that the forces of history are leading to the consideration of larger and larger "households" as the world becomes ever more complex. It is natural, then, that the study of man's "quantification of the value of energy sources" (prices) and his exchange of them with other men (markets) should begin to concern itself with the same problems as the study of ecology, or the general energy exchange between men, other living creatures, and the world environment.

Man was originally considered by ecologists as an unwanted disturbance in the finely balanced interrelationships of plants and animals in an ecosystem such as a forest glade. Now we increasingly see him as a part of a grand ecosystem. In this book we want to suggest what man can do to help maintain a dynamic stability in this universal ecosystem. We do not want to call names, deplore catastrophes, or forecast doom. Instead, we want to discuss some of the economic facts of life and how the science of economics can help us stabilize the environment. Specifically, we will recommend some general plans for government action, in the form of regulation and taxes, together with some appropriate organizational structures. Most importantly, we will suggest a general philosophy of taxation which should help plan ecosystem stabilization actions. Actual problems are usually complex and require individual attention, but a few simple examples are given later to illustrate the methods we are talking about.

The general problem we attack is first that of the nonhuman universe, then the role of man in that universe, featuring the human manipulations that can either stabilize or upset the grand ecosystem. Man's ability to manipulate is great; his need to manipulate intelligently grows as his numbers grow. There is no need to go exhaustively into the entire ecological problem for our present economic purposes. But the future may have some surprises in store for us. So it is best to sketch at least the framework of the big problem: the universe; the earth; and man, with his special abilities and his special problems.

The Universe

The problems of the universe as an ecosystem are so formidable in terms of our ignorance of the facts, and in terms of our ability to do anything about the facts we do know, that we may have to leave to future generations a lot of the worrying about extraterrestrial ecologic problems. But since completeness is the hallmark of ecological arguments nowadays, it is at least amusing to start at the farthest reaches of existence.

If the ecological problem is considered on the most abstract level possible, the processes involved can all be described as the economics of various types of energy exchanges, and the "universe" is really nothing more than a perception on our part of complex energy relations. Matter is energy and energy is matter, as we have learned, to our dismay, during the past century. Energy processes on earth are dependent primarily on the energy from our local star, the sun. Even the small exceptions to this rule (the heat of the earth and in radioactive minerals) can be said to be dependent on past stellar, or perhaps prestellar events. The energy in the stars is presently being produced from nuclear reactions. The conditions for these reactions have been set up by what may be termed, not unfairly, as rather obscure states of heat and gravitational energy. The sources of these obscure states are lost in the recesses of universal history and in the speculations of present-day physical scientists.[1]

This tracing of a natural history of energy is of little direct interest to us. What we want to know is what that energy can do to us and whether or not we can do anything about it.

Starting at the top, it is of interest to us to know the consequences of energy changes in the universe as a whole. Now the structure of the universe as a totality is, relatively speaking, very poorly understood. On the basis of extremely simplified "models," or ways of grouping facts, the universe appears to be a collection of groups of stars, or galaxies, which is about 10^{20} times the size of the earth, or in other words, a very large entity by our scales of measurement. And this collection of galaxies seems to be either expanding, oscillating, or just sitting still.[2] The only thing that appears to be definite is that we in the solar system, or rather we in the local galaxy, or neighborhood of the solar system, are moving away from similar neighborhoods. This, incidently, means that if the universe is "standing still," we are acquiring new neighbors all the time. On the basis of general pessimism, one can well imagine that the expanding universe has some nasty surprises in store for our descendants. Nevertheless, possible consequences are too obscure for present worry on our part to be fruitful.

Coming down from the level of the universe and the Galaxy, we find that our own solar system is a somewhat more rewarding topic of ecological study. It does appear to be possible to guess at the projected future of our sun. The sun is a star, apparently like any other average star, and the stars appear to fall into patterns. Stars are like fires; they have definite amounts of heat and definite flame colors, blue indicating a higher temperature than red. Now, in general, the amount of heat of the fire has no relation to the temperature or level of heat, as one can see by comparing the small, but high temperature blue gas flames of a Bunsen burner with the large, but relatively low temperature flames of red or yellow campfires (or Molotov cocktails). Typical stars, however, seem to be peculiar this way. They appear to show definite relations between the size of the fire and its temperature or color. And, by knowing the size of the stellar fire and how it burns, we can reconstruct its past and predict its future by comparing it with other stars at different stages of burning. By comparing the sun with many other stars, then, we can guess at the future of the sun and of our earth as a member of the solar system. And we are forced to the overwhelmingly probable conclusion that the sun will undergo a dramatic process of change in a future millennial period. This process, due to peculiarities involving the chemistry of hydrogen and helium, will ensure that the sun, in the process of burning up, will grow to enormous size and destroy the solar system as we know it. The action which takes place will be somewhat like that of newspaper set under logs in the fireplace, at the point at which the paper has burned out and the logs begin to flame vigorously. The subsequent, ultimate, final end, that the sun will eventually burn out completely into an insignificant hard ball, is of little (practical) interest.

If we concern ourselves, then, with generations of men and living things in the reasonable future, it is not a matter of fantasy, but one of quite plausible estimation that a home for us must be found outside our present solar system. This type of "environmental transplant" project presents two difficulties. One is in finding a suitable solar system; the other is in getting there.

The probability of finding solar systems on other stars has no well-established answers. Historically, a number of theories have been proposed on the formation of solar systems, whether through the interplay of gravitational and centrifugal forces, through chance encounter of stars, or through the peculiarities of dominant-recessive systems of double stars. A "nebular" hypothesis of formation has been the subject of a great deal of work in recent years, and much progress has been made, for example, in explaining the formation of the observed mixture of elements in the solar system. Also, on another level, the discovery of organic molecules in space has been at least suggestive of life processes outside our system.

But the subject is still speculative, and without a more definite idea of the way in which the solar system was formed, it is difficult to estimate the probability of finding other planetary systems. Various estimates have indeed been made, but for present planning purposes, one must consider that a project for finding a new solar system is a real shot in the dark.

As far as getting to other systems, if they exist in suitable places, we probably will have to accept that travel times will be very long. The fact that there is a theoretical (and experimental) upper limit on travel times has recently been questioned at a very speculative level. However, for serious planning, one must consider that any nearby solar system which is only several human lifetimes away would be quite a find. Problems of the ecology of spaceships, while severe, are deserving of serious study, especially in the optimistic context of twentieth-century insouciance with regard to problems that are "merely technical."

So it seems that we should not worry right now about the overall universe from an ecological standpoint. Strange as it may seem, though, it is quite sensible to worry about the solar system. Nevertheless, there will be sufficient time, on any time scale that we are used to in searching for solutions to technical problems, to plan out projected transfers of population to other solar systems (if any).

Travel from earth to other planets is, of course, quite feasible. The usefulness of other planets as habitats for people and other living creatures appears doubtful according to our present knowledge of the spectacular temperature changes and peculiar atmospheres of Mars and Venus, the two closest, and seemingly most suitable neighbors. The use of planets and satellites (i.e., natural ones) as sources of raw materials for a depleted earth is perhaps a more promising prospect. At any rate, the makeup of the other planets has been investigated during the past decade with the aid of rather large budgets, at least by comparison with the funds allotted to other problems of ecological concern. Therefore, it is possibly of more interest to look now at the poor stepchild of the space race, the earth.

The Earth

It is possible that other parts of the universe may function as escape valves from the technological crisis, at least if we assume a constant or accelerated growth of technological progress in the next few centuries. But in the more immediate future, the present crisis presents a danger of extinction to the human species and other families of life. So we have to specialize our interest in survival to a consideration of the earth as an isolated ecosystem.

The Inorganic Sphere

The active element responsible for most of the energetic processes on the earth is our private star, the sun. Almost all of the processes involving the transfer of energy from one earth system to another depend on this prime source. Cosmic rays and starlight from outer space contribute only very small amounts of energy, as does the heat within the earth itself.

The earth, the target of a small part of the stream of solar energy, is a conglomeration of the less than one hundred elements that occur naturally in the universe. These elements can be specified in detail, but for many purposes the old Greek categorization of water, air, earth, and fire is still adequate. To be sure, nowadays one prefers to talk about hydrosphere, atmosphere, and lithosphere, at least for the first three Greek elements. Regardless of nomenclature, these ancient mixtures of our modern table of chemical elements may be useful abstractions for rough considerations of the form and function of the earth and the life on it.

The earth proper, or the first Greek element, consists of a

9

core of solid material, completely surrounding an interior which is variously solid, plastic, or liquid. This interior is of interest, incidentally, if one worries about the availability of energy in future times. Schemes to tap the geothermal energy source have been recently considered (and carried out on a pilot-plant basis), even by hard-headed, cost-conscious engineers. The composition of the earth is not that much different from the composition of other bodies in our universe. One thing that is basically unusual is that the lighter gases, such as elemental hydrogen and helium (which are so common in stars like the sun), have wandered away from the solid earth. Heavier gases, however, such as oxygen, nitrogen, and carbon dioxide, are held to the earth by gravitational forces and form one of the Greek elements, the air. Of course, the development of volcanic action and life processes has also had a critical role in determining the present constitution of our atmosphere. Oxygen, particularly, appears to be primarily a product of primeval plant respiration.

The third of the Greek elements, water, is perhaps the most dramatic in form in that it occurs in large quantities in several forms. It is a relatively simple compound of probably unspectacular nature, from an objective point of view, but it occurs on the earth in a highly organized form, ice; in a semiorganized form, water; and in a more or less totally disorganized form, water vapor. Water reacts in an interesting way to the changes in the energy flow from the sun that are produced by rotation and revolution of the earth, that is, day-and-night and seasonal changes. These changes unsettle any possible equilibrium of water, ice, and water vapor— water is continually evaporating into the air and then falling back as rain and snow. The odd circumstance that the amount of water available covers only a large part of the earth and not all of the earth means that, in addition, some of this return of water falls on dry land, or earth proper, and is not being balanced by evaporation back into the air. The resulting cycle, beginning with this net transfer of water from sea to land, is an important modifier of the shape of the surfaces of the earth and of the composition of the gases in the atmosphere. This "water dynamism" may be what makes the earth such a peculiarity in the solar system.

The fourth Greek element, fire, does not seem to fit so easily into our modern scheme of nature. We know now, as the Greeks did not, that fire is a kind of process, not a thing in itself, and that it represents the changing of certain elements and their compounds into others, with the resultant radiation of heat and light. But this complex quality of fire might make it a convenient symbol for the manifold interactions of matter in the world. That is, fire can be taken to represent all the different processes of combining materials of the water, air, and earth "elements" into new mixtures. Fire in this sense is the *yang* principal, the life force of the universe, the radiant or motive force which makes change and, hence, life, possible.

We can now form a sort of picture of the inorganic world as a theoretical backdrop of the ecological problems that affect life systems. The sun shines, winds blow, rain falls, and volcanoes and earthquakes reshape the surface of the earth. Against this background of simple, but delicately balanced energy transfer the drama of life begins.

Life Cycles

Life consists of building rather complicated, new compounds out of the primal elements. It is true some simple inorganic processes, such as the rusting of an iron nail, can be characterized as the building of compounds out of elements, and indeed certain lifelike things, such as viruses, share few characteristics with advanced plants and animals. It, nevertheless, appears that the formation of particular kinds of compounds of carbon with other elements characterizes a special category of earthly elements called life, or, if you will, the biosphere. Living things represent a systematic method of taking the energy from the sun and combining it with the passive elements of the earth. Since the irregularities of the earth's surface and the differential heating of air and water produce varying combinations of the passive elements in different places on the earth, life systems have developed a great variety of forms. This proliferation of forms has been aided by the fact that evolution has produced a "quantization" (either-or) phenomenon in the continuation of species: species survive through the overlapping of nondurable individuals, or what we call the phenomenon of individual death and racial immortality. Through death, efficient adaptations of form for all living things have been made possible, and the world as we know it has developed. The early installments of life, of course, were relatively restricted by the supply and form of the earth, air, and water available to them. As life developed, however, life itself made the background environment different. All types of parasitism, epiphytism, symbiosis, and other complex living patterns have become features of the terrestrial life-drama. The interdependence of predators, herbivores, swamp algae, and marsh animals has become a truism of the natural sciences. Indeed, the observation of this interdependence of different types of life systems in ordinary forest and marsh environments was the inspiration for the beginning of the separate science called ecology. It is this close interaction of living and nonliving things in a highly interdependent environment that forms the subject matter of the environmental crisis.

Oddly enough, it is evident that, from the cosmic point of view, there can be no such thing as the ecological crisis. Every type of element, organic and inorganic, finds its niche in the world-scheme. Changing elements in the environmental-scheme, such as a new insecticide or an ice age, can disturb local environment, destroying pelicans or woolly mammoths, as the case may be. The cycle then readjusts; new species are formed by the laws of death and genetics, and the transfer of energy on the earth goes on.

Abstractly, therefore, we can see that the world can be defined as being "quite all right." Naturally, such a Panglossian view of things is more suited to angels than to men. Nevertheless, this fatalistic frame of reference forms a useful background for asking exactly what the ecological crisis consists of. First of all, it can be seen that the crisis exists in the mind of man. Nature can be considered as being objective and disinterested. It follows, then, that there are probably several different ways of considering the crisis, depending on human desires or prejudices. One possible definition of the problem is to require the survival of man in his present state of imperfect civilization. This point of view will be considered in the next section of this chapter and will form the framework of the economic considerations in the rest of the book. But it is not the only definition of the crisis. Many people might require that the ecological configuration of the world should not undergo too great a rate of change. Living species should be preserved and features of the inorganic landscape should be protected, according to this view. But these two different viewpoints of the problem are not necessarily compatible *without compromise* when one looks at actual, possible modifications in human policy to effect certain goals. This discrepancy will also form part of the subject matter below. Finally, one may consider putting only very minimal requirements on the ecology in the world, for example, that the entire biosystem not be erased by a nuclear explosion. Such minimal requirements are, of course, easily fitted (in principle) into more specific demands.

Current concern with ecology is usually centered on more purely human problems, despite the rhetoric of nature worship involved in many environmental discussions. Our particular view here concentrates on the human problem par excellence, that is, on the survival of man. Nevertheless, it is useful to keep in mind the longer view of things in which man is merely a single element having no more a priori importance than any other. But now we will look at the particular role that man plays as both a part of ecology and as a causal agent in the ecological crisis, for man and his generic peculiarities may enter critically into economics at any level. In ecological economics, where familiar market experience may fail us, human factors are often crucial, as we will see later on.

Man

We have described the setting for human existence. Man lives on one of the nine major planets in the solar system, the earth, which is the fifth largest planet and third from the sun (average distance is about 93 million miles); it is slightly flattened at the poles from excessive indulgence in rotation. The earth's age is estimated at approximately 4.5 billion years; however, judging from past experience, this age is subject to change without notice.

The earth is covered with a thin film of matter called life, a film which, according to H. Brown in his book, *The Challenge of Man's Future*,[3] is so thin that its weight can scarely be more than one billionth of the planet which supports it. To be sure, this matter called life is insubstantial and sensitive to an extreme, yet it has existed continuously for most of the earth's history. Man, relatively speaking, has not been around for very long. He made a very late, and probably undramatic appearance some 2 million years ago in this thin living envelope. Worms, scorpions, amphibians, reptiles, winged insects, grasshoppers, birds, and marsupials, to name a few distant relatives, were all on the earth a long, long time before man appeared.

At the beginning of homo sapiens' long climb up the scale

of civilization, man must have been very close to the animal, both in his knowledge of the world about him and in his way of life, i.e., the gap between man and animals must, at one time, have been very small. Man walked on his feet (and hands?), drank the water of the stream, and ate with little ceremony or selection, as do animals. To some extent this is a bothersome, if not downright awkward fact to accept. So in defense of the special qualities of man, some scientists point with pride to his discovery of fire. And more than likely during Pleistocene times in Asia and Africa one of man's close relatives did master fire and its uses. In backward Europe, the technique probably arrived later. And some enterprising paleolithic human groups came to know and take advantage of this discovery.

The early men on earth had few, if any, tools, could not control fire well enough to have it always available, followed animal behavior patterns in relations with their fellow men, and were primarily food gatherers rather than food producers. Also, undoubtedly, man lived in continuous fear of hunger, cold, and attack by other animals.

Primitive man's greatest problem was to secure freedom from want, i.e., to extend his control over his physical environment so that he could diminish the threat of imminent starvation. In order to increase his energy consumption levels man had first to develop skills: to fashion stones to make special weapons, and to build transport devices, and so on. These new skills and innovations helped to increase man's efficiency in hunting, fishing, and killing. But he still lived as a primitive hunter and gatherer of wild fruits and vegetables for all but one percent of his known time on earth.

Somewhere, somehow the great revolution came: the discovery of agriculture and the domestication of animals.[4] At Jericho in the Dead Sea Valley of Palestine; at Jaramo in Iraq and at Tepe Sarab in Iran; at the Chicama and Viru Valleys of northern Peru; at the Bat Cave in New Mexico and

15

in the Nile Valley along the shore of Lake Fayum; and else-where evidence has been uncovered revealing early agricul-tural economies. The "diffusion" of the agricultural revolution began about 7000 B.C., and had more or less replaced the hunting and gathering stage of man's history by 1780 A.D., at which time another revolution, the Industrial Revolution, was already gathering steam. By this time, it had taken a mere 8,000 years or so for man to become primarily a farmer and cattle raiser. It was to take less time to make him into a creature of the industrial age.

Man the Manipulator

Man has observed the world about him and has con-structed in his mind a picture of nature. He has fashioned a system out of what he sees, feels, and smells and has arranged these observations into some semblance of order. Over time, he has tried to understand and control the forces of nature in order to survive. He is *l'homme manipulateur* who, in his un-derstanding of the environment, has, in his way, manipulated it.[5] But his way, associated with the picture of what he saw and felt, was always getting mixed up with other aspects of his life, as we will see later on in this book.

The meaning of one manipulation, the agricultural revolu-tion, was that man had learned to increase and control the supply of plants and animals. It also meant that a major cause of environmental alteration had come into play. Over-grazing, salinization through irrigation, desiccation, and the, at first, unconscious effects of selective breeding on grains and livestock just begin the catalog of changes. The initial step had been taken to control nature's energy sources. The total amount of energy now available for the human species to dispose of increased in proportions inconceivable to the old paleolithic societies. New opportunities were suddenly presented: population growth was no longer constrained; vil-lages sprang up and community life emerged; certain classes

16

and groups of peoples became free of the continuous search for food; and specialization, associated with higher forms of activity, and leisurely speculation became possible. Only ten millennia, plus or minus a few hundred years, separated the beginning of the agricultural revolution from the beginning of the Industrial Revolution, a revolution by which man really began the large scale exploitation of new sources of energy by means of inanimate energy converters.

The term Industrial Revolution may be ambiguous. Here we mean by it the major change that took place in the quality of economic life, in kind and degree, from the development and employment of modern, technologically oriented capital goods. Following the agricultural revolution, and beginning about the time of George III, it was certainly a crucial event in man's history. And this Industrial Revolution today bids fair to conquer the globe, regardless of local race, climate, or topography. Man the manipulator made for himself an industrial revolution and in doing so committed himself to another way of life. Not only is there economic change associated with an industrial revolution, i.e., increases in productivity, but also cultural, political, and diplomatic change, as well as new concentrations of population, and new ways of behaving and thinking.

These changes may appear to be almost basic psychological modifications of man's behavior. In a nutshell, man's attitudes may change: he may become arrogant and agnostic. He may decide that the real God is man—that man is the Manipulator.

Psychology aside, precisely what was the industrial change that man brought about? The changes, as we have seen, were many, but the key change consisted of the substitution of machines for humans and other work animals and the use of inanimate power on a scale never before imagined. The result was a revolution in "energy conversion."

Assume we have a potential form of energy, coal, and

we want to transfer it into a form of energy useful for work. This transfer is only possible if we make use of a converter, e.g., James Watt's famous steam engine. This converter requires an input (i), coal, and, in turn, it generates two kinds of output, which can be designated as useful energy (e_u) and energy of another kind (e_w). e_w is energy loss. In proper mathematical form, then, $i = e_u + e_w$. Although this equation may not rank in complexity with Einstein's, its implications are far reaching. For the input requires the use of natural resources, and in the conversion process we get useful energy plus something else we do not want. Technological improvements of inanimate converters result in less loss or higher levels of technological efficiency, efficiency being defined here as the ratio, e_u/i. The technological march from Watt's engine, with an overall thermal efficiency or ratio of "organized energy" to "unorganized energy" of below 5 percent, has brought us to the modern steam turbine with an efficiency level of 40 percent and to much higher efficiencies for other types of converters.

Plants and animals, including man, are also converters; however, they are quite inefficient, so that in the energy "gear trains" large amounts of working calories are sacrificed. For example, from an energy standpoint (ignoring needs for animal protein) it is more efficient for people to eat corn (see Box 1–1), but in practice corn is fed to cattle that are resected into steaks and hamburgers for carnivorous diets. This loss in energy through the conversion process explains, in part, why less developed countries do not eat meat. And this fact emphasizes how fortunate industrialized societies have been in being able to consume bundles of goods and services far in excess of survival needs.

Man is now considering some of the costs associated with the manipulation of nature's "unlimited bounty" in the production and consumption of extraneous products to the ecosystem. He also needs to realize the consequences of the fact

---

BOX 1–1 / The Human Significance of the Second Law of Thermodynamics

The Chinese have a saying: "Big fish eat little fish; little fish eat bugs; bugs eat mud," and "One hill, one tiger." The ideas behind these two proverbs, when combined, yield a concept, called by biologists the *Pyramid of Protoplasm*. Just what does this mean?

Suppose a man has a pond whose crop is fish. Assume he eats no food other than the fish from the lake; for him, every meal is Friday's main course. Also assume, in this hypothetical example, that it takes about five pounds of fish per day to maintain this man. So in the course of a year it would take nearly a ton of fish to sustain 150 pounds of human protoplasm. In other words, it takes many pounds of fish (the prey) to make one pound of man (the predator) possible.

Assume now that man eats bass, bass feed only on minnows, minnows feed on water fleas, and water fleas are nourished by algae. The second law of thermodynamics, and a little common sense, makes plausible the fact that each time prey-protoplasm is converted to predator-protoplasm there is a loss of organic carbon. In other words: "In any conversion of energy from one form to another there is always a decrease in the amount of useful energy."

One medium-sized pond may support hundreds of bass, thousands of minnows, millions of water fleas, and a mass of algae, but only one man:

"One hill, one tiger."

Source: Garrett Hardin, *Biology, Its Human Implications* (San Francisco: W. H. Freeman and Co., 1954), pp. 673–674.

---

that he is still terribly inefficient in his use of inanimate energy. At the consumer level, nearly half of the original supply of energy, in the course of its use, is dissipated in the form of waste heat.[6] Production and transportation losses, interconversion of fuels, and losses in the conversion of heat to mechanical energy add to the fantastic amounts of energy lost to man. If we add in the fuel elements, such as ash, unburned hydrocarbons, etc., that are residues of the conversion process, the situation is even more critical. The last 100 years may very well be determined by economic historians not to be the era of "high mass consumption," [7] but the "age of waste."

For the long pull, of course, the total amount of energy used may be of concern, especially since some energy is in the form of particular groups of atoms forming elements in our vital resource banks. But we focus on the waste here because the waste not only represents resource depletion, but also contaminates other resources, and this contamination is what concerns us in the short run environmental problem. The waste itself may be unavoidable, but the contamination problem should yield to improvement through better management. We want to explore, later, how economic planning techniques can help us to better resource management.

Nature's "revolution" is nothing more than the compounding of the agricultural and industrial revolutions. And even though nature can afford to be wasteful, genetically speaking, in adjusting to these man-made revolutions, man himself cannot. Therefore, he needs to respond positively to the disturbed ecosystem. Unfortunately, the nature of man dictates rather primitive, if not negative responses to his environment.

THE NAKED APE   To say the least, Desmond Morris' popular book, *The Naked Ape*,[8] is a stimulating speculation on the behavior of man. Man's sex, rearing, exploration, fighting, feeding, and comfort habits are all, he maintains, related to more humble origins. These related behavioral patterns associated with primates and animals further down the evolutionary chain have been speculated on by others. For example, Robert Ardrey, in his book, *African Genesis*,[9] has accumulated vast amounts of information pointing to the intriguing theory that homo sapiens developed from carnivorous predatory killer apes, and man's age-old affinity for war and weapons is a natural result of this inherited animal instinct. In his second book, *The Territorial Imperative*, [10] he tries to show how human behavior has been shaped, but not determined by environment and experience and how it is a consequence not of human choice, but of evolutionary inheritance. Does man behave like the Mongolian gerbil, a charming little animal that marks out its territory with its sebaceous gland and whose earth-boring activities would make a prairie dog envious? Is it the nature of man to stake out a territory [11] and even to bore into the ground? Maybe so.

Certainly, all three works undoubtedly suffer from various shortcomings in the empirical establishment of their arguments. We believe, however, there is a grain of truth in their main argument: that man is indeed a biological specimen, a part of nature, and, therefore, tied in many ways to his evolutionary inheritance. A thesis based on this premise would conclude that the ecological problem confronting man today is, in part, directly related to his biological past; that is, it is his evolutionary nature to change the environment in a deleterious way, to waste resources, and, in doing so, to further evolutionary change.

This evolutionary change is made possible through another kind of waste, genetic waste. Garrett Hardin, in his book, *Nature and Man's Fate*,[12] has convincingly shown that genetic waste, or redundancy of genes, is an important factor in determining change; whenever external circumstances alter the "fitness" of the species in his environment, the species is instantly prepared to alter to meet the new demands through mutation. The needs of any given emergency are met through the adaptability conferred on the organism by ever present genetic waste. In essence, this means that man the manipulator, in influencing his environment and creating large volumes of wastes, forces the "grand lottery" to meet changing circumstances. Man, of course, is not infinitely adaptable; the new environment cannot suddenly be too different from the old, else man simply becomes extinct. Man may truly be a

21

naked ape, but he has adorned himself with some interesting garments that reflect his past history and basic psychological drives.

THE APE IN SHEEP'S CLOTHING Sigmund Freud wrote in his book, *Civilization and Its Discontents*,[13] that civilization is only made possible by the individual's renouncement of his instinctive life. The instinctive life of man is one of aggression and egoistic self-satisfaction. And Freud asserts that the whole structure of culture has been designed to put prohibitions and curbs on man; he sees man's place in the world in terms of the never-ending conflict of the claim of the individual for freedom and the demands of society. These demands and man's reaction to these demands are different in each society. In other words, the nature of man as a picker of berries, a harvester of nuts, and a forager of tubers and the seeds of plants may have been far different from the nature of man who now pushes a cart to gather cans of berries, frozen foods, and cream-filled chocolates from the shelves. But the eternal conflict between freedom and necessity remains. Physiologists may prefer to view this conflict in terms of responses of the hypothalamus and the cortex to cultural stresses. In any event, is it really in the nature of man to be maladjusted to his environment?

Yes. At least in some sense. In the confusion of the exponential changes brought on by man himself and the biological mandates of nature, man is torn between two worlds. The bonds to the atavistic world are not easily broken, so he must bring psychological mechanisms of defense into play in dealing with interpersonal and interenvironmental problems. These defenses, of course, have changed over time, and the normalcy of an individual has been judged by how well he adjusts to the natural and societal mandates of his era.

In any case, in the action, reaction, and interaction of men with themselves, as well as with the environment, and the production and the exchange of goods and services, we see man as a biological and societal phenomenon. Whatever his state of mind, it is of little concern to nature. The biological behavioral patterns are given. Societal behavioral patterns are variable and, hopefully, subject to change. But the last few hundred years have witnessed the development of strong and powerful trends that will make adjustments towards a stabilized ecosystem difficult. The foundations for the free enterprise type of society have been well laid.[14] In religion and in parallel developments of commerce, finance, and industry, habits and institutions have been molded to produce a general philosophy or state of mind which has directed this society to its present economic and environmental state. Resolution of pressing problems in the economic and the environmental world can come about if, and only if a new philosophy prevails. Until that time, man's relationship to man and man's relationship to nature has resulted, and will continue to result in an inadequately controlled ecosystem.

The Human Brain as an Ecological Appendage

We have just viewed the ape and the ape in sheep's clothing as individuals, groups and organizations in their associations with one another and with nonhuman conditions and events. These associations are unusually complex because they involve the human brain, an ecological appendage or specialized outgrowth which lies both inside and outside the ecological interactions and provides a perspective on the behavior of human organisms towards the ecosystem. It is the ecological appendage that directs our attention to various kinds of phenomena. These include, among others: the psychological behavior of persons (singly and in groups of various kinds); the undertakings or strategies designed to achieve envisioned goals or purposes; and the outcomes or the operational results of such undertakings, including outcomes that are unintended as well as those that are intended. With respect to all of these phenomena, explanations and predictions are likely to reflect some idea of environment and some hypothesis of relationship between the person or group in the context of surrounding conditions and events.

Also, the intellectual power of this ecological appendage may be thought of as a power of "appropriate selection," so that when it is given difficult problems it provides correct answers: this would be the behavioral equivalent of high intelligence. If this power of "appropriate selection" can be amplified, it seems to follow that intellectual power, like physical power, can be amplified. In this sense, the brain provides a basis for a *cybernetic* system.

CYBERNETICS AND THE WORLD OF CLAY   Norbert Wiener created the name cybernetics—"the science of control and communication, in the animal and the machine." [15] The word cybernetics is derived from Greek roots meaning "steersman" and is associated with coordination, regulation, and control. These communication facets provide the foundations for assessing and resolving the problem of making a stable, viable environment, or what we may call the "ecostabilization" problem.

It is the delicate nature of self-adjusting systems, referred to also as homeostatic (a word coined by Walter B. Cannon) or cybernetic systems, that concerns us, and the implications of imposing constraints on these systems in order to return to the norm. Systems can only return to the norm when the response (feedback) is negative with respect to the stimulus that evokes it. When the impressed charge is positive, the response charge is negative, and vice versa. Negative feedback is characteristic of all self-adjusting systems, which, as the reader will see, includes the mechanisms of the economic systems developed in this book.

It is possible that a negative feedback system may break down and be replaced by positive feedback, which leads to a runaway process. For example, if body temperature rises to around 107° F, the normal negative feedback system breaks down and is replaced by positive feedback. Increased temperature causes faster metabolism, which causes higher temperatures, which . . . , and so on. A similar positive feedback occurs when the body temperature falls 10° below normal, resulting in a runaway process in the opposite direction. In general, self-maintaining systems are self-adjusting only within limits. Once outside these limits, pathological possibilities exist (see Chapter 11), and in the Darwinian cybernetic system, in the presence of environmental change, secular or long-range change towards another norm is the rule.[16]

Cybernetics deals with the way things are organized, not with the nature of things in themselves. It concerns itself with machines in the most general sense, that is, with complex interacting mechanisms. The mechanism involved may be a thing of steel and germanium, such as a computer, or a complicated set of inorganic and organic molecules, such as a woodland glade. And the cybernetic system may be analyzed, even if it is still a dream in an engineer's or economist's brain; it need not exist in the actual world.

The concepts of this very general field of study can then be used to investigate all types of human and natural systems.[17] By combining the human brain, the environment, and the world of clay, it is possible to represent the most diverse type of system as a problem in cybernetics.

To say the least, the ecological problem is a very complex system. Cybernetics allows us to deal with this complexity. It provides effective methods for the study and control of systems that are intrinsically extremely complex. It will do this by first marking out what is achievable, and then providing generalized strategies of demonstrable value that can be used uniformly in a variety of special economic-ecological cases. In this way, it offers the hope of providing the essential methods by which to attack the pollutionals of our economy that, at present, appear to be defeating us by their intrinsic complexity. The electronic computer is, of course, the means by which many dreams of cybernetics can be made into reality. But the computer, of course, is just a big, fast adding machine in itself. It serves as a tool for man the cyberneticist to try to help solve the environmental feedback problems that perplex us today. The use of computers is implied in much of the planning we discuss below, just as writing implies pen or pencil. But the pen and pencil do not do our thinking for us and neither do computers (at least, not much, not yet).

We do not pretend to offer perfected methods of attacking the environmental problem, but we do attempt to suggest a foundation for man and his world of clay to start in the right direction of "ecostabilization" (or stabilization of ecosystems). Of course, this problem would be simplified, though not solved, if man would slow down on his breeding habits.

Too Many Men?

According to United Nations demographers, world population is doubling every thirty-five years. Should the trend continue, which it will not, in another 700 years there would, theoretically, be one person for every square foot of land presently available on the earth's surface. Of course, the present key importance of the process of photosynthesis, to name one constraint, would prevent a "standing room only" world. But it is the short-run implications of this trend that will seriously affect the economic, social, and natural environment.

The present rate of population growth appears to be more rapid than it has been in any earlier time in history. United Nations figures show that the population of the world in the year 1 A.D. was approximately a quarter of a billion people. By the year 1650, it had doubled; by 1800, it had doubled again; by the early 1900's, it had doubled again; and by 1970, it had doubled once more (which presently makes for an annual increase of some 70 million). The average annual growth rate for the period 1950–1965, 1.8 percent, is substantially higher than earlier periods: 1750–1800, 0.4 percent;

1800–1850, 0.5 percent; 1850–1900, 0.5 percent; 1900–1950, 0.8 percent. What all these statistics mean (and they have been presented so often) is the fact that there are over 3.5 billion people pressing on the resources and environment of this planet and more are coming.

Of course, as with all data, we must be careful of the figures before 1800 in that the first real census did not take place until 1801 in England, under the direction of John Rickman; in Italy, 1861; and in China, 1950. To be sure, there were earlier censuses, for example, there was a very famous census in the time of Augustus Caesar. The question present-day demographers have in relation to the old censuses is: Were women and children counted, or were only men counted for conscription purposes into the army? Earlier censuses in China were based on such things as salt consumption and mail delivery. Later ones were too. More meaningful methodologies are employed today in the estimation of world population figures, and the margin of error is something less (plus or minus 200 million) than pre-1900 estimates. The United Nations projections to the year 2000 and 2050 of 6.8 and 25 billion people are, on the other hand, the results of some rather sophisticated demographic models. Remembering that the *ceteris paribus* (other things remaining the same) clause has been employed, these models predict that we can expect a population in the year 2050 of approximately 25 billion; but *ceteris paribus* is almost certainly not a realistic assumption.

Paddock and Paddock in their book, *Famine 1975! America's Decision, Who Will Survive?*,[18] have viewed some consequences of these short-term population growth figures. They maintain, and support empirically, that the world's ability to support higher levels of population by 1975, especially those less developed economies with insufficient technologies to maintain larger numbers of people, will result in the removal by death of tens of millions of people by starvation.

This is the Malthusian specter in every sense, acting as a particular form of Nature's harsh wisdom.

In the late 1700's, Robert Malthus was asked by his father, Daniel Malthus, to comment on a book written by William Godwin entitled *Political Justice,* a book which gave promise of a distant future in which "there would no longer be a handful of rich and a multitude of poor. . . . There will be no war, no crime, no administration of justice, as it is called, and no government. Besides this there will be no disease, anguish, melancholy, or resentment." [19] Godwin's vision, like many others of the Romantic Era, suggests, irresistibly, the dreams of the flower children of today, where, like their Utopian predecessors, they attribute men's suffering primarily to corrupt government, greedy employers, bad laws, and so forth.[20] Thomas Malthus (probably due to the generation gap) disagreed with his father's sympathetic attitude towards Godwin's book and certain reforms advocated by Condorcet. To convince his father, he wrote his objections, published in 1778, in a work called *An Essay on the Principle of Population as It Affects the Future Improvements of Society, with Remarks on the Speculations of Mr. Godwin, M. Condorcet, and Other Writers.*

What did Malthus have to say in this work that has become a classic reference in our culture? One thing is for sure, he could not accept the views of the Utopians, and he feared that the "need for food" and "the passion between the sexes" would result in anything but an imaginary land of everlasting peace and plenty. He showed, with empirical evidence from the American colonies primarily (the first census of the United States was taken in 1790), that population increases geometrically (1–2–4–8–16–32–64), whereas the subsistence from land appears to increase only arithmetically (1–2–3–4–5–6–7). (See Box 1–2.) That is, the exponential increases in population are not supported by similar increases in the productivity of land. The inevitable result, he wrote, was to be a world of "misery" and "vice." To be sure, he

BOX 1–2 / The Law of Diminishing Returns

Malthus' views depend directly on the law of diminishing returns, i.e., the production of food *tends not* to keep up with the geometric progression rate of growth of population. (1–2–4–8–16–32–64) versus (1–2–3–4–5–6–7–). Why the gap?

With some existing state of production techniques, as more and more of one productive resource (say labor) is used with a fixed quantity of another productive resource (say land), the extra additions of output will eventually get smaller and smaller. In other words, the law of diminishing returns refers to successively lower extra outputs gained from adding equal increments of a variable input to a constant amount of a fixed input. The table presented illustrates the principle. Additional workers have less to work with. For each additional mouth there will be two more hands, but no more land.

MAN YEARS OF LABOR	TOTAL PRODUCT (BUSHELS WHEAT)	EXTRA OUTPUT ADDED BY ADDITIONAL UNIT OF LABOR
0	0	
		2,000
1	2,000	
		1,000
2	3,000	
		500
3	3,500	
		300
4	3,800	
		100
5	3,900	

Obviously, if proportional changes of labor and land led to proportional changes in the output of wheat, and if labor yielded increasing extra outputs, the world's wheat could be grown in a flower pot, if the pot were small enough.

See also Appendix A, and Chapter 4, note 5.

noted, there were certain positive checks, factors operating chiefly as determiners of the death rate, which had kept populations down in the past. These positive checks included wars, "vicious customs with respect to women," famines, "great cities," plagues, "unwholesome manufactures," and "luxury." There was also a second kind of check on the past growth of population, which Malthus called a "prudential" check, reducing the birth rate. This preventive check, he maintained, was the postponement of marriage. Later, in his second edition (1803), with the magnificent title, *An Essay on the Principle of Population or A View of Its Past and Present Effects on Human Happiness with an Inquiry into Our Prospects Respecting the Future Removal or Mitigation of the Evils Which It Occasions*, he was to advocate "late marriage," and "abstinence within marriage" to keep population from growing too fast. Thus, the thin thread of "moral restraint" was the only hope to prevent mankind from drowning under

---

BOX 1–3 / Economic and Social Determinants of Birth and Death Rates

A broad brush approach of fertility and mortality patterns, as they are affected by economic and social forces, suggests the following:

Fertility—1. Other things being equal, age, or specific birth rates, tends to vary directly with per capita income in the long run.

2. The urbanization process tends to reduce birth rates in the long run.

3. There is a negative correlation of birth rates with the level of education.

4. To a certain extent, overpopulation tends to generate its own antidote.

Mortality—1. Other things being equal, there exists a negative long-run association between death rates and economic conditions.

2. Urbanization and industrialization seem to play a significant direct role in the reduction of mortality.

3. Mortality is negatively associated with differentials in medical care.

4. There is a negative partial correlation between the rate of growth of real per capita income and death rates.

Human catastrophes associated with positive population checks have resulted in relative stable populations. However, as societies experience an industrial revolution there are "demographic explosions." Why?

Consider the following rates for a pre-industrial society. The starting point is a high birth rate (35–50 per thousand) and a high death rate (normally 30–40 per thousand up to 100–300–500 per thousand). With an industrial revolution the high, recurrent death peaks tend to disappear. Under the impact of progress in medicine and sanitation, as well as improvements in the diet, the normal death rate undergoes a downward movement. The birth rate eventually follows but at a time lag. The movements of this rate are subject to a complex interplay of heterogenous forces, some of which are mentioned in the text. The point is that the gap is widened between births and deaths, adding new fuel to the explosive growth.

We now pride ourselves on controlling disease, and we have this humanitarian urge to give medical assistance to societies which are still basically agricultural. In Ceylon, we provided DDT to wipe out the malarial mosquito, and deaths fell from 22 to 12 per thousand in 7 years. This decline in mortality was not accompanied by any measurable change in fertility—the crude birth rate remains at over 40 per thousand!

Source: Irma Adelman, "An Econometric Analysis of Population Growth," *The American Economic Review* (June 1963): 314–339. For more detail see Philip M. Hausen and Otis Dudley Duncan, *The Study of Population* (Chicago: The University of Chicago Press, 1966).

its own weight. Malthus duly noted that it was not the nature of man to refrain from such pleasurable activities; that the dictates of nature on early attachments to one woman and the "moral restraints" of abstinence or marrying late do not "prevail much among the male part of society." Parson Malthus, of course, would never have advocated birth control, and the use of his name by Aldous Huxley in *Brave New World* for a prophylactic belt carried by the Greek-lettered members of that society would have caused him considerable concern.

Population movements before the Industrial Revolution have shown high birth and high death rates (see Box 1–3).

And it is an economic axiom that in any country that has not experienced an industrial revolution, the birth rate is, on the average, between thirty-five to fifty-five per thousand per year; the death rate is between thirty to forty per thousand per year. The gap between the birth rate and the death rate is, to say the least, an interesting one. For example, take Adam and Eve some 700 thousand years ago, and ignore Archbishop James Ussher's biblical beginnings and Isaac Newton's historical chronology.[21] Assume that the gap between births and deaths over this period of time remains between .5 and 1 percent. If this rate had been maintained, the world today would be a ball of living flesh expanding faster than the speed of light (with Einstein's permission, of course). Why hasn't this happened? The positive checks about which Malthus wrote have come into play. Famines, epidemics, wars, and plagues have appreciably affected the death rate, increasing it temporarily far above the natural birth rate. Over the centuries these human catastrophes have resulted in a relatively stable population.

It should be obvious that the agricultural revolution and the industrial revolution have created productivity techniques that support higher and higher levels of population. It is possible that another revolution is in the making, i.e., the atomic revolution, after which unlimited supplies of energy, theoretically, could support an unlimited supply of people, at least if the world could be air-conditioned.[22] But objections to such a mass sea of flesh appear appropriate considering the serious density levels in many lands throughout this world today. It needs only to be mentioned that the compounding of men has, so far, resulted in a compounding of the ecological problem. Perhaps the solution to these problems is a pandemic plague. In any event, nature is taking the matter under serious consideration.

2

ECONOMIC ANALYSIS: MEDICATION FOR ECOSYSTEMS

It is hard to deny that economics plays a crucial role in the problem of the troubled world ecological cycle. Phrased in the broadest possible terms, the importance of work, money and the exchange of goods and services is too well-known to all of us. We ignore the realities of the marketplace only at our peril. We can protest this reality and wish that the world were not so, but even the most fervent revolutionary must ultimately bend to the gross facts of economic life.

On the other hand, economics in the narrow sense of a social science is regarded quite differently by some people. The value of using the sophisticated techniques of economic science to plan our world is sometimes a matter of controversy, and distrust of economics as a science does exist in various segments of our society. Even if such distrust were confined to the younger and more radical elements of society, economists would be concerned with trying to correct the misunderstandings behind such hostile attitudes. But even in the ranks of the not-so-silent majority, there is, undoubtedly, little basic trust of economic science, except, of course, when it happens to back up selfish interests.

Naturally, the situation is different in certain elite segments of the establishment of our society. There is, for example, a steady movement in scholarly fields toward the casual acceptance of such modern econometric (i.e., mathematical-economic) techniques as "linear programming" and "macroeconomic modeling." This acceptance and understanding will undoubtedly filter down eventually from economic scientists to experts in various specialties and from them to planners and the general public. So the eventual use of advanced tools of economic analysis in the treatment of the problems of modern planning, and especially the problems of man and ecology, may be taken as assured.

But this eventual use of economic analysis as a mere part of the solution to ecological problems is not in itself an adequate response to our needs. Disturbed ecosystems require, by their very nature, a highly unified approach to planning. Any benefits from an intelligent use of economic analysis, in such circumstances, can be easily lost in the shuffle of conflicting political and social considerations. This is especially true because the ecological issue has become an emotional one: survival has become a watchword, and claims that a particular solution is "too expensive" generate public resentment and journalistic rage. Such reactions are, of course, quite understandable, since people do not ordinarily accept economics as a basis for all life decisions. Neither is such a one-sided "economics only" basis suggested here. The ecological problem requires a unified scientific approach based on sound analytic methods.

So what we propose here is not merely the use of the tools of economic analysis, but the creation of a unified system for ecological policy decisions. This unified system would have economic analysis as its hard core. That is, instead of dealing with a potpourri of different value systems (sociological, political, economic, etc.), value choices would be made on the basis of economic "decision models," or ways of planning, generalized to include factors which (to the noneconomist, at least) appear quite different from conventional dollars-and-cents considerations.

Economics is suited to this central, hard core role in an ecological planning model. The basic content of economics is a reflection of some key characteristics of human nature and social organization. What we have to do is to see how to enlarge this content to enable planners of the present and future to treat all the major facets of the troubled ecological cycle.

The Worldly Earth

Economics deals with the world as it is, rather than with the world as it should be. In this, economics, of course, shares a common outlook with the physical sciences and with the other social sciences. We know that research during the past century has greatly aided our understanding of how closely economics is dependent on human psychology and on the sociology of human organizations. However, economics still suffers from a kind of class discrimination, as documented in Robert L. Heilbroner's book, *The Worldly Philosophers*,[1] in which he describes the early economists as investigators into the then unfashionable gross realities of life. Economics has come a long way toward social respectability as a field of knowledge since the days of Adam Smith. Leaders in government and business have come to accept, or at least give real consideration to the analyses by leading economists of national economic trends and goals. But some influential groups in our society still often tend to feel that economic analysis is somehow related to a negation of idealism and progress (in the sense of humanistic development). We believe that this feeling is rather wide of the mark and that a study of the works of the great economists would reveal it to be more prejudice than fact.

Prejudice of this kind is of little concern for scientific work and for many practical applications, since the economist usually works either with fellow researchers who are sufficiently sophisticated to understand the place of modern economics in the totality of social science, or with businessmen or planners who are fully committed to purely worldly values. But in the context of the ecological crisis, such misconceptions about economics are rather serious because recent popular movements concerned with ecology have attracted the attention of unusual mixes of people. Among these various people are many to whom economics is equated with the demon called "materialism." In this diabolic role, economics is supposed to be irrelevant for the planning of improved ecological management, or worse, economics is assigned the major blame for the ecological crisis itself. So it is important to see how a properly generalized economics can represent basic facts about human existence and how the analytic tools of economic science can be naturally useful in the stabilization and improvement of disturbed ecosystems.

First, we might like to anticipate some of the criticisms of the William Godwins (see Chapter 1) of today, for the student revolutionaries and the self-styled "freaks" of the younger generation have managed to set up enclaves, mental or physical, within our civilization in which they try to deny the universality of some of the psychological axioms of economics. Against acquisitiveness, they set sharing; against complex division of labor to produce surpluses, they advocate the reduction of desires and the reintegration of labor with the needs of personal satisfaction. They bill themselves as harbingers of the future, that is, as prophets of a future utopia of sufficiency and economic humanism.

So these elements challenge us to defend the relevancy of our concern with economics. For what good is economics, if a viable utopia, without the striving and financial ambitions of the present, is in the cards for the future? Now no one can answer for the future, so such a utopia may indeed come. And even if utopia never arrives, dreams are the stuff not only we ourselves are made on, but also the stuff of future fruitful social experiment. But utopianism should not become mere obscurantism, because obscurantism is sterile and sterility is death, for people and for society. So, utopias or not, it is important to ask precise questions about economics and the future. We should try to see why economics is not only important in today's world, but why it will probably continue to play a role in the society of the near future, even in societies of rather different, perhaps utopian character. Starting at one beginning point, we can ask about the future of value, the concept so central to economics. How will value, or worth, appear in the context of ecosystems of the coming century?

The Uses of Lethargy and Impatience

In understanding the why of economics and particularly the rationale of value, economists usually stress the role of patterns of human wants and the fact of scarcity of goods. But for our special purpose of changing a perhaps grudging acceptance of the facts of economics into a full appreciation of the latent power and beneficent possibilities of economic analysis, it is better to look at these concepts of value in a somewhat different way, in terms of the production factors and their costs in future societies. There are two reasons for this. First, the pattern of human wants, except for the most basic needs, tends to be changeable according to various cultural patterns and psychological conditionings. This fact means that the antieconomist, or humanistic idealist, is able to view patterns of wants as being manipulable by future, more perfect governments or societies. In fact, some of the most important support for increased awareness of ecological problems has come from radical movements among young adults who have rejected existing want patterns.

The other reason for approaching the problem of value

from a somewhat new direction is that the history of technological progress has led some people to question the future
importance of scarcity of goods as a market determinant. It
may seem obvious to some of us that goods must remain
scarce, in spite of technological changes, because of the evident continuing scarcity of resources in all future times. Still,
it might be true that goods will become exceedingly inexpen-
sive in terms of values per average labor hour. Such a cheap-
ening tendency might make it unnecessary to take detailed
economic analysis into account in evaluating future world
ecosystems. That is, goods might become so plentiful and material wants so easily satisfied for the citizen of the future that
the concern now shown for economic matters might be prin-
cipally transferred to esthetic (or, perhaps possibly, erotic)
matters.

Unfortunately (or fortunately, depending on the point of
view), such a devaluation of goods to very low exchange levels seems unlikely, despite apparent evidence to the contrary
provided by events of the past two centuries of industrial expansion. We can argue that this historical trend probably will
not go on by considering, not scarcity and want themselves,
but, instead, the factor inputs which go into the production of
goods: labor, capital, land, and management.

Labor is one of the most interesting of these inputs. The
economist talks about the disutility of labor, or reluctance to
work. It is important to examine why this disutility or unwillingness of people to work will remain important regardless of
future technological changes. In the first place, it seems clear
that people will always prefer certain types of work or pas-
time activities to other work or activities. So relative values
for goods seem a permanent feature of the world. There is, indeed, evidence of this in the nature of the public struggle
against pollution and other ecological problems. For example,
with certain praiseworthy exceptions, people tend to demon-
strate against, rather than pick up discarded tin (or alumi-

num) cans. Even recent community disposal efforts have often been primarily demonstrations to promote publicity. It is probably not too bold to suggest that demonstrations are "preferred activities," at least in relation to tin can collection. In general terms, this tendency toward activity preference may be termed a recognition of the eternal role of relative lethargy in economic life and human existence. It follows, then, that various products (assuming for simplicity a homogeneous labor supply) will continue to have values relative to the price of labor invested in them (*ceteris paribus*); and those products produced by relatively undesired activities will still tend to be more relatively valuable, utopia or no.

Even if the persistence of relative value in future utopias is admitted, however, it might still be conceivable that in a future world so much excess time will be available that people will be willing to contribute their labor for what is, in effect, a zero price. In fact, leisure could become so common and leisure time so extended in scope that work, in general, might be positively valued as an activity. Conceivably, in this case one would not have to work to derive exchange tokens (money) for particular goods. One might, instead, with the help of advanced automated facilities, personally manufacture all desired goods.

Naturally, we think that this case, which indeed seems somewhat fantastic by present-day standards but which is imaginable in the future, will probably not be realized. For there enters another fact of existence, the mortality of man, or in more everyday terms, the vice (or virtue) of impatience. When a person has surplus labor time available, he may not be willing to wait the length of time needed to manufacture the goods himself. So, he will demand sufficient price for his labor at relatively preferred work activities so that he will be able to make immediate exchanges for the goods he wants. This reasoning holds true even if the "manufacture" consists of no more than pressing computer control switches at a communal automatic shoe (or candy, or rocket) factory; for we may guess that the sense of impatience will grow keener as absolute labor times grow less. At least we know that we now endure with impatience several hours of travel from New York to San Francisco. We can well imagine that a forty-niner would not sympathize with us at all!

In conclusion, then, it seems that impatience and lethargy are contributing human factors in the preservation of value (in the economic sense) in future utopian (or anti-utopian) societies.

Capital and Other Demons

Naturally, other factors besides labor contribute to the input value of goods. The critical role of capital as a factor in producing goods at ever-increasing rates seems secure well into the predictable future. The unit price of an average capital good may change in unforeseen ways in the future. With the tendency toward higher levels of real output (more and more goods), capital goods, such as die presses, may become cheaper and cheaper in relation to the cost of labor hours. An incidental political consequence of this trend may be a decrease in negative emotions produced by the words "capital" or "capitalists," emotions recently made fashionable again by neo-anarchist movements among the youth of today. Probably capital will then become, for future ecosystems, more important per physical unit but less important in total value (and, incidently, more neutral in emotional tone).

We should also take into account current trends toward expanding the realm of what we call "capital." Economists have noted that the process of education can be considered to create "human capital," [2] for education represents an investment of time and money, both in explicit costs and in earnings foregone, that makes possible increased future earnings. So in his economic role the educated man is just as much a capital good as is a lathe or a milking machine.

Unfortunately, this educational capital is itself subject to depreciation, as the course material painfully ingested in youth fades, with disuse, from the memory. So future economic planners may well share the concern of educators with this obsolescence problem and with related cures, such as programs of continuing education.

All this is not to say that man is a machine, or that education is not a source of personal satisfaction in itself. Indeed, economics could, in principle, deal with this satisfaction as a consumer good of as yet undetermined value. But even though only a part of the value of education could be so handled in money terms, this part may, nevertheless, play an exceedingly important role in future economic calculations. Already the inclusion of educational capital has been useful in "balancing the books" in economic calculations: differences in (effective) educational levels from country to country go far toward explaining troublesome discrepancies in observed income per worker in different cases. And economic analysis can, undoubtedly, help educators and governmental administrators to make more rational choices in spending money on schools and colleges. Presumably, as a society we would like to maximize the return on such investments, but there is little agreement or hard information on this "return," for example, from the three main functions of higher education: instruction, research, and discovery of talent. Great gains in income are commonly attributed to a college education, but these calculations are notorious for their neglect of hidden costs, such as income foregone by students during their time at college. It might well be that elementary or secondary education deserves relatively more support than colleges, from an economic point of view.

Such questions of efficiency are accompanied by questions of equity. Does university training contribute to more social justice through the raising of poor students to the affluent ranks of the professional? Does the educated elite of one generation breed the educated elite of the next, and is higher education, therefore, a regressive factor? Answering such questions on a scientific basis is as yet a hope and dream for the future. But we see that the educational problem will provide much work for economists in that future, and economic science will play a critical role as the role of education becomes more and more dominant as civilization grows ever more complex.

The role of land is somewhat ambiguous. Certainly the land now used by man on the earth itself will become relatively more scarce. Countering that tendency, however, will be the utilization of much remote and inhospitable territory on earth through improved communications and climate control. An economic role for land on other planets and in other solar systems is, of course, more speculative, at least in the near future. But in the long run, if man survives, Archimedes' "place to stand" may become a scarce commodity indeed, and Martian real estate may become an economic good listed in the "Properties, for Sale" section of the Sunday newspaper. The landlord may regain his feudal dominance of society, or alternately, ownership of all land by the state or the single tax scheme of Henry George may be adopted to rescue this scarce factor from private misuse. At any rate, land as an economic entity will be very much with us.

Of course, in all of this we have looked at only the passive, topological use of land. If we include the natural resources of the earth (and the solar system?) in this factor, we see that scarcity, and, therefore, economic accountability, is very much in the picture for the future. Known oil, gas, and coal reserves are measured in amounts that may well be exhausted within the next few centuries. Expected increasing scarcities of metals such as molybdenum and vanadium may be even more critical to mankind, since the use of nuclear energy may alleviate some of the need for fossil fuels (not, of course, the need for them as sources of hydrocarbons for synthetic materials). Obvious partial answers to this scarcity of resources lie in recycling durable materials, and the economics of recycling is in its infancy. So here too there is plenty of work for the economist of the future.

The last factor category, that of management, will undoubtedly retain its key position in producing value. It may be, of course, that future civilizations will become static societies in which management or entrepreneurship will become relatively unimportant, at least as a percentage of value added to products (in the form of profit or managerial salaries). Nevertheless, experience during the past few centuries seems to suggest increasing roles for management. Such an entrepreneurial outcome appears at least plausible on the basis of present knowledge. Naturally, the entrepreneur of the future may be called a commissar (or perhaps guru).

So it appears plausible that the keystone of economics, the presence of a significant finite value for goods and services, is destined to remain a fact of life into the foreseeable future. In particular, based on very general and respectable human traits, labor inputs or the amount of sweat involved should still command a substantial price in future societies. Such considerations are, of course, of no surprise to economists. They, naturally, will disagree on exact projections of future economic structure, but economists and planners may find that they have to be ready to use some such arguments on economic value for counter criticism (often implicit) from present-day radical Marxist and neo-anarchist circles.

Given such a probable continuing importance for the concept of value in future ecosystems, the role of markets and other economic mechanisms must be looked into in more detail as they apply to the ecological crisis.

The Marketplace

One generally envisions a market as a place (or an area) where things are bought and sold—a hubbub of activity with entrepreneurs calling out their wares and a pickpocket or two. The economist, however, interprets the market in a broader sense (and more formally); i.e., he views the market as a set of pressures exerted by buyers and sellers for a total of related transactions. These transactions take place in an enormous variety of markets: markets for crude oil or copper, which differ from markets for tickets to *Prince Igor* (or *Kismet*); or markets for cattle, which may not seem like the market for pig iron.

The market in economics can be looked at as a control mechanism for the productive activities of man. Since man and his brain now function as a control (admittedly imperfect and ill-informed) acting to modify the ecological cycle, the market becomes a key link in the ecostabilization problem.

The operations of the marketplace are intricate and interrelated, something akin to the workings of the human body, in which widely varied viscera function to serve each other without any conscious determination. Yet there is an important difference in that there are no control directing systems in the marketplace, such as the brain, and no single set of purposes, such as our conscious personal desires. Each buyer and each seller enters the marketplace with his own set of purposes; and, to quote from Adam Smith's panoramic book, in entertaining their own "self love" they are moved as if by an "invisible hand" in the direction "which is most agreeable to the interest of the whole society." [3]

In a highly simplified way of looking at it, the market system operates by the exchange of labor, in its many forms, for money. Business firms use these services to produce other goods and services which they, in turn, exchange for money.

In this "circular flow" goods and services move from firms to consumers, and a reverse flow of money from consumers to firms occurs. As long as the economy is neither expanding nor contracting, these flows will be constant through time. The connecting exchange link between the two flows is the price of goods and services.

Price Mechanisms

Every society, in producing and consuming goods and services, will have some level of technological information, a fund of resources provided by nature, and desires or wants of varying intensities for a myriad of goods and services. The Herculean task of the economic system is to bring together these technologies, resources, and desires, and to do so efficiently. Any society must somehow resolve the economic problems of what goods should be produced, what quantities should be produced, how should the goods be produced, and for whom should the goods be produced. In ant or bee colonies, all such problems, even those involving an extraordinarily elaborate cooperative division of labor, are solved automatically by means of so-called "biological instincts." An omnipotent dictator, benevolent or malevolent, may well resolve the task of the economic system. In the enterprise-type economy ("capitalist free enterprise economy") the tasks of what, how, and for whom are primarily resolved by a system of prices. If resources were unlimited, if human desires were fully satisfied, and if an infinite amount of every good could be produced, the what to produce, how, and for whom would not be a problem. There would then be no economic goods, i.e., no goods that are relatively scarce. All goods would be "free goods," and there would be little need, incidentally, for economists.

But people want more in total than society can produce, and they may not want many products that could be produced.[4] Therefore, priorities or goals must be established. Society not

45

only needs to determine what to produce, but it must also determine in what quantity—a very difficult problem when considering the interconnections between all the many commodities wanted by a modern society, e.g., the proportion of the number of automobiles to barrels of crude oil. In the enterprise system, priorities (or values) are established by the consumer who "votes" with his dollars for goods and services which he desires to have produced. His control over production is limited by the size of his income, i.e., desire is simply not enough. While most Americans probably want a yacht or a fur coat, only a few have the income to support these desires. The vote of the consumer (consumer sovereignty) is a characteristic of an enterprise type of economy in which the intensity of the desire of the consumer for goods and services controls the composition of output. Assuming entrepreneurs are motivated by the desire for profit, they will devote their productive efforts (if they can cover costs) to those goods and services most urgently desired, as evidenced by dollar votes.

In addition, society must allocate its resources so that in a particular state of technology it can obtain as much as possible of the output wanted. It is important to remember that this allocation of resources is not just a technological question. Certainly, there are many ways of making every product, and engineers may sometimes prefer one way to another for irrational reasons. But even at best, technological information is not enough to make an efficient choice. Efficiency in this context is an economic concept, not an engineering one, and the economic system has to decide how to allocate the resources efficiently. Prices govern the allocation of resources in the enterprise system. Entrepreneurs whose products are selling at high prices will be in a position to bid for productive services used to produce products less urgently wanted, so reallocating resources to the production of those products that consumers want more.

Another factor is that the total output produced by the economic system must be distributed to the consumers of the society. And this division is a function of income, that is, the purchasing power of the individual determines how much he gets of the goods and services produced by the society. Money income, in turn, depends on the ownership of resources (which is usually primarily labor), and, of course, people can act to change this income and ownership pattern. Through the acquisition of skills, the price paid for the ownership of labor resources may be increased. People may also increase their contributions to society by begging, borrowing, or stealing their neighbor's goods or talents. But independent of individual differences, it is a characteristic of the enterprise system to reward and to punish monetarily the members of the society whose contribution to that society is great or small.

There is an additional function of the enterprise system which we particularly need to mention here, that is, it must be able to change with time, not only to meet the needs of a growing population and the adoption of new technologies, but also to respond to disturbed ecosystems. It must be capable of producing new commodities and of coping with exogenous forces, such as war. Prices in the enterprise system are the mechanism for coping with change and growth. Should the Douglas fir beetle (*Dendroctonus pseudotsugae*), wind, fire, or a hyperactive lumber company destroy standing timber in the Western United States, a rise in lumber prices compels the use of substitutes and forces buyers to economize in their use, and compels lumber companies to reforest rapidly. So superior environmental technology may elbow its way in by either lowering or raising costs.

Forces in a market are pressures exerted by people with opposite interests. Sellers compete with other sellers for the best sale. Buyers compete with other buyers for the best buy. Competition between sellers as a group and competition between buyers as a group exerts pressure that puts people "on their mettle" and tends to force them to serve others better than if there were fewer pressures. More important, the rivalry helps direct efforts and scarce resources to their best use; it induces us to add to the things we and others want most. Such is the interaction of the two forces, supply and demand, in the enterprise marketplace. Since, these two terms, supply and demand, represent, by far, the largest part of the economists' vocabulary (and that of lay people too), we have presented a more formal review in Appendix A.

Capitalism as Destiny

The problem of survival—of the essential production and distribution of goods and services—for our society is solved everyday in this country not by custom or command, but by the free interaction of profit-seeking men bound together only by the market itself. This type of system is called capitalism. Its critics are well-known. In any event, take your pick of what is wrong in the world and you will undoubtedly find someone has blamed it on capitalism. Other economic systems have also had their critics. Isn't it wonderful? Nevertheless, one very serious issue with which we and other members of this society are concerned is whether the present system is capable of improvement; whether, for example, it can solve the ecological problem.

The population of this society, through its consumption habits, supported by a fecund industrial complex, is violently pressing on the ecological environment. As Professor Galbraith writes in *The Affluent Society,*

The family which takes its mauve and cerise, air-conditioned, power-steered and power-braked automobile out for a tour passes through cities that are badly paved, made hideous by litter, blighted buildings, billboards and posts for wires that should long since have been put underground. They pass on into a countryside that has been rendered invisible by commercial art. . . . They picnic on exquisitely packaged food from a portable icebox by a polluted stream and go on to spend the night at a park which is a menace to public health and morals. Just before dozing off on an air mattress, beneath a nylon tent, amid the stench of decaying refuse, they may reflect vaguely on the curious unevenness of their blessings. Is this, indeed, the American genius? [5]

We are plunged, whether we like it or not, into a struggle for survival. Whether we survive is going to depend on the present system and its ability to solve the ecological problem. This problem, to a large degree, is economic in nature, and the world of fact requires adjustment of production and consumption patterns. How and at what cost these patterns are changed will be determined in the market. And you had jolly well better believe the ecological crisis is coming to the marketplace.

All of this, of course, assumes that the capitalistic marketplace is a constant of existence. But is capitalism really necessary? In many parts of the world, "free market" is just the name of an obsolete institution in a foreign fossil society. If socialism is the wave of the future, why bother working within the present system?

To be sure, there are all kinds of guesses as to how institutions like capitalism will develop or disintegrate as time goes on, just as there are all kinds of socialism, from Tito to Mao. It is impossible to predict what will happen, so perhaps we had better be prepared for all eventualities. But we can defend a preoccupation with capitalism with several reasons: (1) Capitalism exists now and is very much alive and living in the Western world (and in much of the Eastern). (2) Our free market economy contains a very important government sector. So in considering capitalism we have to include most of the issues that face a socialist planner. (3) It is an amusing fact that socialism, at least in Yugoslavia and the Soviet Union, is showing signs of becoming somewhat "capitalistic" (shades of Marx and Engels!).

On the last point, Yugoslavia is perhaps a special case in that the necessity for the revision of socialist institutions (e.g., factories owned by workers) is more or less explicitly recog-

49

nized. In Russia, the story is rather different. Socialism remains unchanged, but capitalist methods are of increasing importance for "accounting purposes." The Soviet Union has long had an economy based on *khozraschët*, or economic accountability. This means that state enterprises act as if they were private, in the sense (limited, to be sure) that relations with other state enterprises are based on buying and selling on a monetary basis. And within the past decade, there has been growing interest in establishing a true quasi-market economy in which prices are negotiated between buyer and seller. Profit-sharing, true economic rents, and other free-market ideas have been put forward.[6] Many of these ideas were inspired by the suggestions of Ye. G. Liberman, who analyzed the key difficulties in the Soviet economy as being due to arbitrary output norms and arbitrary factor prices or quotas.[7] He recommended that incentive payments to managers be based on profit (return) on capital invested.

One swallow does not make a summer, and the future course of "profit Marxism" is not clear. But these theoretical economic developments in the Soviet Union might give pause to those who, like Khrushchev, or perhaps Eldridge Cleaver, are ready to "bury us."

Of course, we can't begin to look into the details of a comparison between the two major economic systems as just a footnote to our concern with environmental problems in a free enterprise economy. Still, we might venture to say, speculatively, that capitalism is not all bad, even according to its opponents. Economists assure us that a perfectly functioning competitive system (an artificial beast) does assure an optimum allocation of productive resources in the economic system (if not an optimum distribution of goods),[8] and it is nice to know that our economic system works so well, even if in an idealized theoretical form. Further, as far as the environment is concerned, the difficulties are certainly not exclusively capitalistic.

Capitalism does hurt the environment. But socialism, or what passes for it, has also been guilty of severe environmental disturbances.[9] Russians have smog and water pollution too. So as in many instances in practical life, maybe ideology does not solve everything. We can't depend on Karl Marx to save us from pollution. So let's concentrate on improving our own system and solving the environmental degradation problem in our own world as it is.

The Uses of Coercion

In our discussion of the marketplace we were viewing a competitive world—a world in which the number of buyers and sellers was sufficient to promote equilibrating adjustments within the marketplace and, in the process, allocate the scarce resources of the society to their best use. In the real world, pockets of industrial concentration have evolved, and depending on the degree of their concentration, they may appreciably influence or control the market mechanism (prices). This phenomenon exists when one, or a few firms in any given industry behave in such fashion so as to misallocate resources thereby providing the society with something less than it could have. Also, the government enters the market as a buyer of goods and services, and its monopsonistic (sole buyer) power is well known (ask any aerospace company), and through legislation (like tariffs) and other actions (like taxes), it either forces or prevents adjustments in the market, e.g., by establishing maximum or minimum prices, which may prevent an equilibrium adjustment in the market and result in surpluses or shortages of the goods or services.

Market Imperfections

Operations of market systems have been hampered to a degree by the nature of the structure, conduct, and performance of American industry. Over the years, through mergers, acquisitions, or internal funding, various firms have evolved into giant corporations, and associated with this "bigness" is the ability to control or force the market mechanism. However, we should not get caught up in the "conventional wisdom," to borrow a term from Galbraith, that the United States economy is strongly monopolistic. If we act the part of the skeptic, there simply is no body of knowledge which can tell us this is so. Of course, we are entitled to be skeptical about the skepticism itself.

The tens of thousands of studies [10] undertaken by economists and government agencies to theorize about and perform analyses of the levels of concentration, and the economic implications of a few firms eventually producing most of the output for any given industry suggest that the U.S. economy is certainly not the classical economist's dream world. The question is: Are market imperfections getting worse or better? Certainly, the 200 or so large corporations are still growing, and mergers continue. The fact is we (economists) are not sure. Seventy-five years ago there was considerable concentration in some industries, but our statistics do suggest that the growth of concentration has leveled out in the last few years. This is, in part, due to antitrust legislation and the proper enforcement of this legislation in more recent times. The literature on unfair trade practices, to say the least, is savory. In addition, it may be argued that the business environment has placed some constraints on the continued tendency towards higher levels of concentration; to grow unrestrictedly larger is just as much an invitation to disaster to a business as it was to the dinosaur. The key, of course, is not whether a firm or industry is competitive in nature, but rather whether the firm or firms within an industry are, to borrow a term from the American economist J. M. Clark, "workable," [11] that is, whether they are producing the goods and services which the members of this society desire. If we are going to criticize

American industry and its present conduct in contributing to ecological instabilities, then we must also ask what the likelihood is that the present industrial system will solve the ecological problem. Is the system still "workable" under these new demands on it? Does it behoove American society to make the system more competitive or should society place legislative constraints on the operations of the present structure? Is there some alternative approach which can best resolve the ecological problem? We are not only confronted with the classical goal of efficiency in the economy, but also with the foremost aspects of efficiency in the allocation of resources so as not to appreciably harm the ecosystem.

It is the economic nature of an imperfectly structured industrial system to misallocate resources; in a competitively structured world resources are optimally allocated. However, both an imperfectly structured system and a more competitively structured system may influence and disturb the ecosystem, which makes for a real bed of thorns for the economist.

If that is the case, then why all this talk about market imperfections? Simply put, the environmental policy maker must take into account the market structure before any meaningful policy can be initiated. Imposing a constraint on the system will affect an imperfectly structured system in one way and a competitively structured system in another. Much of our later analysis takes this distinction of market structures into account.

Levels of concentration vary from industry to industry. One sector of the economy may be more competitive than others. This is so because there are conditions which appreciably contribute to market imperfections, and, hence, the number of rivals. Some firms, given the nature of the product they produce, must expand their operations in order to enjoy certain economies or lower costs of operations. For example, to produce the quantity of steel the members of this society desire requires the operation of large and elaborate facilities. The diminution of these facilities could possibly result in higher costs.

Superior quality entrepreneurship is another factor which is not enjoyed by all firms. These qualities not only lead to lower costs for the individual firm, but make it difficult for other firms to compete successfully against such decision makers. Superior entrepreneurial abilities, like those of Ford, Sloan, Rockefeller, and Carnegie, are generally so scarce that, when available, they may lead to concentration in any industry. Some firms have developed because of their ownership of indispensable resources. Mineral rights of one form or another should suffice as an example. Exclusive franchises protect the operations of some firms in their production process, say through patents, and prevent others from employing these licensed techniques. Money capital is, of course, the sine qua non in promoting levels of industrial concentration and the associated market imperfections. That is, it enables a firm to undertake activities that smaller or less established firms simply cannot finance. Money power can purchase economies of scale, quality entrepreneurship, indispensable resources, and franchises in their many forms. One or all of these factors (or some combination) contribute to higher levels of concentration and may hinder market operations by presenting formidable barriers to the entry of new firms within a given industry.

With classical competitive conditions more or less nonexistent, and in those industries in which a few firms account for the largest share of the market, we would expect, and do in fact find attempts to differentiate the products being produced. This in turn, influences the nature of consumer demand. The possibilities of differentiation are enormous—differentiation by the way the producer's output is packaged or marked, by the way the product is styled, by the conditions for sale, etc. Heavy product differentiation occurs in the consumer goods industry, especially in the durable goods industry where a Chevrolet is an automobile but a Rolls Royce is a Rolls Royce. It is this distinction or differentiation that affects demands of consumers for the product in the market.

We add: Even those manufacturing industries that sell to other producers are not free of differentiation. For example, major users of coal make an exact appraisal of the different grades of coal from well-known regions. Giant electrical utility companies, for example, readily measure this difference.

Consumer demand represents the other side of the coin in supporting market imperfections. It is in the marketplace that the consumer evaluates different brands and establishes various preferences or levels of satisfaction for some bundle of goods and services. Here the product fills no technical function but satisfies different personal needs, both mystical and physical. Some products cannot be easily differentiated, such as wood, steel, and wheat, and distinctions are not easily measured by the individual consumer, although there have been attempts to differentiate even these products for him, e.g., the Weyerhauser Timber Company now stamps its products with little green trees (but a $2'' \times 4''$ is still a $2'' \times 4''$, even though its size seems to shrink over the years). The promotion of these products in all sectors to consumers and to producers by differentiation may encourage higher levels of demand, which in turn may accelerate ecological instability. But Weyerhauser, in employing better technology, may be a better protector of the environment than the many small lumber companies. The little green trees, then, may take on new and significant meaning to the buyer.

We will study more closely below the nature of monopoly and oligopoly (domination of a market by one or several companies). We call such institutions "serpents" because their actions often make invalid a good deal of the reasoning of classical market economics. Besides this theoretical difficulty, monopolists have often been blamed in practice for a good portion of the injustices of society. As we shall see, whether they are unjustly maligned or not is rather beside the point. Certainly, like death and taxes, they are with us, but should we learn to coexist gracefully with "The Detroit Three" and all their kind, accepting what it appears we are powerless to change? Who should adapt—the consumer or the producers? One thing is for certain, if the environmental position of these firms remains fixed, like the dinosaur, they will die. They must adjust. The question, then, becomes whether or not their rate of adjustment is fast enough.

MONOPOLIES AND OLIGOPOLIES: SERPENTS IN EDEN   Industries that are characterized by the existence of monopolies and oligopolies, as contrasted with the typical competitive situation, display some interesting behavioral patterns. These marketplace behavioral patterns of this type of firm arise within and because of the total environment of the firm itself, other firms and industries, and the consumer public.

In a perfectly competitive world, in terms of its market structure, there would be low levels of concentration, no product differentiation, and insignificant barriers to entry. In this environment, the firm could not appreciably influence the price of the product it sells and could make no choice with regard to the product it sells. There would be no advertising budget and rivals would not influence its operating decisions. But the presence of market imperfections changes all that. For example, the pure monopolist is literally the industry, in and of himself, and he has an independent hand in determining price and output, or the amount produced. The monopolist will have some optimal advertising budget and some quality level of his product. He may pursue ends other than maximizing profit goals, e.g., increasing his market, creating or maintaining a desirable public image, fulfilling social responsibilities, and so on. Whatever price and output decision the monopolist makes will not correlate much with the way he deals with the external environment.

In an oligopolistic situation the possible patterns of conduct become much more complex. That is, the few firms within the industry are cognizant of the fact that they must take into account the impact their policies (pricing, output, and other) will have on their rivals' decisions. For example, when one firm cuts its prices, it considers the possibility that its rivals will also cut their prices. This is a distinctive element of market conduct as compared with both pure competition and monopoly, where individual sellers react only to impersonal market forces. In an oligopoly they react to one another. John von Neumann and Oskar Morgenstern have shown in their book, *The Theory of Games and Economic Behavior,* that when mutual interdependence exists the sellers do not take into account just the effect of their actions on the total market, but the effect of their actions on one another.[12] The theory of games has contributed little to our knowledge about oligopolistic behavior [13] simply because the rules of the game keep changing, and even if all the participants know the rules at any given time, they do not necessarily adhere to them. However, the analogy drawn by von Neumann and Morgenstern between game theory and oligopoly tells us two valuable things about oligopolistic markets and ecosystems. First, a distinction can be made between a firm's action in initiating its own change towards environmental policies, and its response to some environmental action initiated by one of its rivals. Second, mutual interdependence is a matter of degree, and when more than two firms are present in an industry no one firm can be sure how a given rival will react to its environmental policies or whether it will react at all.

Are monopolies necessary and do they fulfill a natural role in society? In several sectors of the American society natural monopolies do exist, especially in public utilities and transportation. Firms that provide electric power, natural gas, telephone service, water, and bus transportation in a city are examples of public utilities. Their monopolistic positions are sanctioned by the public authority at the state and federal level and operate under detailed regulations governing business practices. Oligopoly also exists, and there are several sectors (steel, automobiles, cement, cigarettes) in the U.S. economy in which a few sellers do take account of the effect of their own output and price decisions on those of their competitors. Their presence may be a natural phenomenon, possibly a necessary consequence of the planning requirements of advanced technology, as argued by J. K. Galbraith in *The New Industrial State*, but this point is open to debate. At the other end of the spectrum, the theoretical pure and perfect competitive market does not exist, but there are industries that probably come close. A widely used example is the farming industry in which no one farm produces enough to put a dent in the total corn, wheat, or cotton crop.

Many "intermediate" firms also exist. In particular, a great number of firms do not have the powerful position of an unregulated monopoly or well-coordinated oligopoly, but neither do they obey the rules of open competition. Such firms operate under conditions that are not monopolistic or oligopolistic in the sense of being able to set prices by fiat or by consultation with other firms, and yet they do not deal with the homogeneous products and the "externally-determined" prices of the truly competitive sectors. They have a monopoly of a sort, in that they have an identifiable product of their own or at least a group of customers more or less loyal to the firm or to the product brand name. But this monopoly can be broken if the firm raises its prices too high or if the competitor brings out an improved product or a clever advertising campaign.

According to many economists, this type of firm is very common, and this kind of "monopolistic competition" [14] is extremely visible on the American scene. Indeed, this type of firm sometimes makes a public spectacle of itself by indulging in apparently senseless, destructive price-cutting campaigns. For this type of firm is apt to precipitate disastrous price wars if it lowers prices to attract more customers. To avoid this fate, competition in this sector is often expressed instead in improving products to attract new customers. This product improvement may be improvement in reality, or in fantasy, whence extensive research on new products and large advertising expenditures.

Now the general behavior of the "monopolistic competitor" may be rather complex to analyze, but we want to concentrate on the effect of the market on the environmental problem. And for this purpose, the "monopolistically competitive" firms can be thought of as sharing both some of the characteristics of monopoly power and some of the features of the competitive jungle. In particular, they may deserve special treatment in the tax policy we suggest in Chapter 9.

The operations of these intermediate firms do feature one phenomenon that is of special interest for the environment: the necessity for producing new products. As we have noted from the gasoline additive F-310 of Standard Oil and the non-phosphate detergents recently introduced on the market, these new products can incorporate environmental concern as, so to speak, part of the product itself. So this product-competitive part of the market does have a special contribution to make to antipollution drives: as a result of "monopolistic competition," the consumer can sometimes buy a cleaner environment.

All these kinds of market conduct draw a great deal of attention (sometimes anguished) in our society. When steel prices are increased, when a new secret ingredient is added to a particular brand of gasoline, when American businessmen are indicted for conspiring to fix prices, or when the government sets price guidelines, market conduct becomes front page news. For our purposes we wish to know whether or not this market conduct will have an important effect in the resolution of the ecological problem, because some types of conduct will have important implications for the performance of the marketplace in resolving environmental crises. To cite one example, assume, as in the example above, a major oil company introduces a new smog-free fuel and argues, through various media, that every user of this new fuel will be contributing to clean air. Keeping the example simple, assume other firms in the oil industry all follow the single responsive policy of lowering the price per gallon of the old type of smog-producing fuel by five or ten cents. We leave it to the reader to reason out the net short- and long-run economic/ecological effects. Hint: both answers will probably be wrong.

Anyway, if we looked at a big industrial map of our country today (including Hawaii, Alaska, and the territories), we could see an enormously dynamic and complex system of production: products being generated both in competitive and less-competitive environments. In order to produce the aggregate bundle of real goods (television sets, aircraft carriers, automobiles, golf clubs, Saturn rockets, and so on) one would see lines of factor movements, all directed by the structure of prices, from primary sectors; mines, forests, and farms; to steel, lumber and textile processing centers. These materials, in turn, become inputs for other sectors and industries for other outputs, and so on down the line until the product is purchased for consumption by one of four consuming sectors (personal consumption, investment, government, foreign trade). The net effect of a comparatively small number of materials developed in the primary sectors is a remarkable bundle of goods, all developed by an "enterprise" system. This production process proliferates into an amazing variety and incredible amount of waste.

Production is the transformation of inputs. Things are bought by a firm, and with the various technological and procedural recipes it has at its disposal, things are produced which it sells in the marketplace. However, there may be inputs that the firm does not pay for, such as salt water to cool nuclear generating facilities. And there may be output that the firm does not sell in the market, e.g., smoke. This phenomenon is of considerable interest to us and is treated at more length in Chapter 4. But whatever the elements of an industry's structure happen to be, they can all disturb the ecosystem in the production process.

Saving You from Yourself

Were it not for consumption patterns and, to a large degree, the habits of the consumer, many ecological problems would not now be entering the marketplace. It is the nature of consumers to demand those goods and services that are scarce, and should the society somehow fully satisfy the desires of individual consumers, they would more than likely shift to another scarce good or service. Until individual consumers practice austerity or asceticism, industry will continue to support their desires for these goods and services, and, in doing so, *ceteris paribus*, will keep on disturbing the ecological system.

The economic and social cost of the deterioration in environmental quality has increased. These costs are a function of consumer demand and use of goods and services, and of the production process used in producing these goods and services. An improvement in environmental quality may come about by altering either the demand or the supply side of the economic equation. For example, it would be possible to tax those goods that contribute most to disturbing the ecological system, so that a decrease in the supply would result in higher prices and less of these goods being produced. Conversely, we can tax the individual consumer or penalize him for purchasing those goods that disturb the ecosystem. The question is, of course, how much should we tax the consumer and how much should we tax the supplier? There is a myriad of alternatives and there are many adjustments that can be made. The question is, which one (or combination of several) is best? Let us proceed slowly before attempting to answer.

Planning and Power

The ability we now have in this century to analyze men as economic beings, as they influence and are influenced by prices and wants in the marketplace, is a genuine triumph of the human intellect. Like many really important achievements, the understanding of the equilibrium tendencies of pricing patterns according to utilities (see Chapter 4) and scarcities is basically a simple concept. If we were engaged in just a search for knowledge for its own sake, we could profitably follow this simple concept, as it is modified by the circumstances and details of human interactions with the environment, and continue the development of a complete science of human economic behavior. But from the point of view of the ecological crisis, this positive side of economics is only a part of what is necessary. To correct the deficiencies in present ecosystems, the market must not only be understood, it must be modified or, frankly speaking, manipulated to achieve desired ends. We will call this manipulation the "planning process." We are all familiar with planning in such areas as the development of adequate water resources for large cities, the carrying out of rural electrification, and the open market sale of securities by the government to affect money supply. The popularity of the phrase "mixed economy," as applied to our enterprise system in the United States, implies the acceptance of a degree of planning, even in capitalistic contexts. The nature of economic planning in general is a subject unto itself; in this book we merely indicate how planning may be adapted to particular problems involved in the stabilization of ecosystems.

Some people object to government planning on principle. Some object on practical grounds. But ever since the Keynesian revolution in economic thinking, it has been as difficult to ignore the crucial role of government in the economy as it would be to ignore the bull in the china shop. It is by no means obvious that the government planner must take over the note of protector of the environment. But in terms of power, the government is indeed a highly plausible place to look for help. We adopt that choice here and discuss it further later on in Chapter 12.

The way to do ecosystem planning through the use of economic analysis will form the subject matter of several of the next chapters. We will glance briefly at the *motivations, means,* and *ends* which have to be considered in the formulation of appropriate ecosystem stabilization plans.

Motivations

Our ultimate motivation in developing planning possibilities for the solution of the ecological crisis is often popularly expressed in one dramatic word, survival. The survival of mankind and various other living species is often said to be in danger, if present trends continue. Certainly, the need for long-term survival must be a basic driving force behind our concern here. How does this motivation fit in with other existing human motivations, particularly those germane to the subject matter of economics?

In economics, one generally talks about a drive for increased profits, implying by this that the market should operate in such a way as to maximize the total amount of goods and services available to everyone. We have all been taught to believe that this is a good thing. Even if the distribution of goods (social justice) has been at fault, the basic idea of continually producing more and more gasoline lawnmowers and polyethelene baggies has been accepted as the way to go.

Naturally, professional economists are aware of the subtle and sometimes arbitrary assumptions inherent in such a point of view, but the public is often not. Consequently, some people think that the ecological crisis has arrived just because of a basic conflict between economics and survival, with the drive for profits having led to the danger of destruction of the world as we know it. Air travel, for example, has grown considerably; more people travel to Europe now than formerly traveled to the next state. But the means of travel threaten us with air, noise and, as in the case of the SST, even glass pollution. The danger to quality of life or to life itself from any one such result of economic development may be trivial. But we seem to be in danger of being overwhelmed by an accumulation of trivial contributions to the streams of environmental pollutants, and these pollutants are produced as the by-products of a primarily market economy.

We maintain that the viewpoint that casts market economics in the role of a dyed-in-the-wool environmental villain, while it contains elements of truth, is misleading and barren for purposes of a practical approach to survival. Fortunately, alternate ways out are available, and it is possible to see that there need not be conflict between survival and profit if profit is correctly treated to include the long-range view of mankind.

An example may help us to see what the long-range view means. During the first World War the demand for foodstuffs induced farmers to introduce wheat into what had been short-grass country in the Western United States, between the Rocky Mountains and the beginning of the longer-grass prairies in the extended Mississippi Basin.[15] These new lands produced good crops of wheat and, therefore, high profits for farmers during the war and well into the twenties. But it turned out that the farmers during those years had been especially lucky; most years were of sufficient wetness to produce successful crops. Scientists knew all along that the natural ecology of the region involved fluctuations of wet and dry years. When a succession of dry years came during the thirties, the famous "dustbowl" was produced, impoverishing vast numbers of farm families and contributing to the overall economic depression. Obviously, by ignoring what was already known about the regional ecology, agriculturists had produced short-term profits at the expense of what were undoubtedly severe long-term losses. Profit defined in the long-term sense would have been a useful planning concept in this case.

The situation is more complex when questions are raised about natural esthetics and the preservation of nonutilitarian species of animals and plants. These questions will be considered in more depth later in our examination of proper criteria to be used in a definition of overall social (economic) "profit."

Means

Once motivations have been selected, plans for the alleviation of socio-economic problems can be designed to bring the world into harmony with the motivations of human beings as, hopefully, expressed through the planner. There are various means or methods by which the planner can then affect the world. One of the most straightforward methods is by fiat: regulation or legislation. In effect, the power of the state can be used to threaten people with punishment if they do not behave in consonance with the particular plan and goals, in this case, the stabilization and improvement of existing ecosystems. Punishment may include imprisonment or monetary fines. The concept of monetary fines or penalties shades insensibly into the payment of civil damages for ecological malfeasances, such as air pollution or careless deforestation. Monetary damages, in turn, melt gradually into the general concept of taxation. Public bodies can apply a tax penalty to undesirable practices which are believed to injure the ecosystem. These taxes, of course, are economically quite similar to damages or fines. But fines and other criminal penalties also imply social disapproval, or in somewhat different terms, moral obloquy. It is obvious, then, that another permutation on the implementation of planning can be through the use of moral disgrace alone. So, propaganda or education can also

65

be used as a means of carrying out the planner's work. Again, propaganda and education seem to have somewhat different connotations. We might think of propaganda as smacking of the coercion of a Stalin or a Hitler, while education is a noble, pure, and liberating force. But reality may not be so simple. Propaganda, oddly enough, may be thought of as a noncoercive measure, in that people are persuaded that their true long-term profit, in the economic way of thinking, lies through modified action on their behalf. For example, it is a conservation of one's energy to throw away an aluminum beer can along the highway. But careful propaganda can convince a citizen that this action will lead to a long-term deprivation in the sphere of esthetics. Education, on the other hand, while partaking of this instructive quality, also implies hidden coercion. That is, those educated on an issue are usually defined as being morally more correct than those uneducated. There are, then, various interesting psychological interweavings of the manipulative devices available in the various main categories of means (or weapons) in the ecological struggle. So, generally, economic and noneconomic means are, in practice, not distinct entities but occur in the form of complex mixtures in most policy decisions.

We could, of course, carry the level of abstraction one step further, by calling criminal punishments an "infinite tax," in the sense that the loss of reputation and freedom could be defined as being beyond money compensation. Similarly, we could carry out an "economification" of propaganda and education, by looking at them as being similar sometimes to infinite taxes and sometimes to methods of clarifying accounting procedures. These procedures could be related to showing people through public instruction where the truly long-term profits for society lie.

Such generalizations may seem rather strange for ordinary use, but in the construction of mathematically optimized plans, such increases in levels of abstraction may be helpful in getting at the basic choices involved in practical dollars-and-cents planning.

Ends

In a sense, the end or goal of the planning process is already contained in the motivations themselves. A better world that gives more pleasure, comfort, and security to the human population might serve as at least a very general goal. Naturally, such words as "pleasure" have to be defined in a wide sense, to embrace many of the disturbing aspects of ecological disequilibrium that are found presently in the world. An exploration of such problems will be a large part of the concern of this book.

In another sense, however, it is useful to imagine what the end, or end result of the planning process would be in tangible terms; what existing institutions would change, and what new institutions would have to be added? Without doubt, the planning envisioned would require the necessary evil of a large bureaucratic apparatus. Much as some of us might wish to return to simpler eras or to some utopian future, such a requirement seems inevitable when one considers the vast growth of agencies that have been formed in past centuries to administer the regulation of public goods and certain private goods. This tendency may have bad features, but it is there, all the same. So we may as well consider such a growth of public administration as a necessary evil. At least if this necessity is faced squarely, one can do one's best to ensure that new agencies are formed and the existing agencies are rendered as efficient as possible, perhaps with the help of continuing research (and, hopefully, a big breakthrough) in the methods of applied social science.

We would also expect to see the government acting more and more through the market, somewhat as the Federal government now acts in selling securities and in the placement of government research and development contracts. Again, the

67

inevitability of such a larger role by government should suggest due concern with the structuring of the institutions involved, and especially with checks on bureaucratic power. Of course, we are not talking merely about national tendencies here. Since ecology is an international problem, it is to be expected that new institutions must be created at the international level. As experience over the last few years shows, difficulties at the world level will be similar in scope to those at the national level and will be doubled in spades in actual complexity.

3

THE WORTH OF VALUE: A BRIEF INTRODUCTION TO THE PROBLEM OF CRITERIA

Deciding on values is one of the hardest jobs of the planner. Without believable criteria for making decisions, the whole planning process is crippled before it even begins. The problem of relating the usual financial criteria employed by the economist with the fuzzier criteria treated by other social sciences and by humanist thinkers will be discussed more fully in the next few chapters. Since the subject is complex and controversial, it may pay here to take a brief synoptic view of criteria, or the problem of values. Chapters 4 and 5 will deal more fully with this problem area.

The Weakness of Virtue

One way of solving problems is for the planner to decide what is right and then encourage everyone else to act accordingly. This method may be called the method of "virtuous conduct." Phrased in such simple terms, the fallacies in such a point of view may seem obvious. But it is still worth considering here because the utilization of virtue as a means to an end in ecological problems has, in fact, been a popular approach in recent years, even if such attempts to change behavior by exhortation might sometimes be presented thinly disguised as science or economics. Often, of course, any practical method of environmental protection has been difficult to apply because ordinary citizens feel helpless to change their environment. So in the absence of concrete ways to change ecologically destructive behavior by business, government, or citizens, such stop-gap measures as publicity campaigns by conservation groups and investigating committees, and expressions of disapproval in private discussions and in newspapers, at least serve to relieve deeply felt frustrations. Such campaigns and discussions are useful both in leading to regulatory and countereconomic actions and in generating direct public relations pressure on offending corporations or individuals.

While we can accept the general usefulness of such activities, we cannot rely on them to solve the ecostabilization problem. We must do more. For one thing, publicity campaigns are notoriously dependent on what ideas happen to be currently fashionable. Indeed, ever since pollution has become a cause célèbre cynics have been noting that pollution is replacing poverty as the fad of the moment. Such criticisms can create an atmosphere of opinion which will eventually cripple any practical progress in antipollution measures. Also, the effectiveness of propaganda campaigns, formal or informal, may depend critically on the charisma or expressiveness of the particular people involved. Small changes in personnel can then make disturbingly large changes in the effectiveness of the movement. Society, like bureaucracy, is effectively built on the supposition that for policies to be viable they must be executed on a sustained basis by personnel who range in capability from the good to the mediocre. But who can depend on continually getting good, in the sense of vigorous and capable people to lead action groups? The recent history of some civil rights groups illustrates the problem. Campaigns and action groups are always helpful and are fine altogether for problems that are not so critical. But the ecological issue is too important to be subjected to the vagaries of chance inherent in the hortatory approach, better known as the "call to virtue" by aroused citizen groups.

We must also realize that the publicity type of approach must be modified or supplemented for the simple reason that it has often proved ineffective in practice. Take the case of smog in Los Angeles Basin. After twenty-five years it is still there and spreading yearly.[1] It is true that public outcry and interest by local government officials have produced many solid results in the form of legislation in prescribing controls on burning, industrial emissions, etc. Finally, after twenty-five years relatively severe standards on automobile emissions were set recently by the state of California. Those concerned about air pollution have been opposed by automobile manufacturers and oil refining companies, who understandably emphasize research progress while giving practical difficulties and failures less advertisement. Nevertheless, recent contentions by Ralph Nader and his group to the effect that corporations have been disgracefully remiss in fighting pollution, must seem at least plausible to all of us on the basis of our knowledge of human nature, disregarding for the moment any judgment with regard to the accuracy or completeness of the facts brought forward by Nader. In a word, some progress can undoubtedly be made by public pressure leading to legislation for the correction of ecological dysfunctions. But in view of the uncertainties and inefficiencies inherent in this approach, it would be foolish to rely on it when questions of survival are at stake.

All this means that just adopting a particular solution to an ecological problem and then publicizing this solution to secure either informal or formal regulated compliance is a dangerously ineffective approach. But if the right-or-wrong approach to questions of value is taken, this approach is the only feasible one. So other, more subtle approaches to worth and value must be taken. The use of economic methods of valuation, even in approximate form, will make possible the use of more effective tools to change current dangerous tendencies in our human ecosystems, but these methods of evaluation may have to be widened to reflect the complexity of the pollution problem, as we will see later.

Money as a Measure

If right and wrong are insufficient methods of evaluating the factors in an ecological crisis situation, what are other feasible, and possibly more useful methods? In measuring choices between objects in everyday life, money is certainly a straightforward measure of value. We see that money, as it relates to the usefulness of things to us, may be used as at least a partial method of placing values on various complicated planning options, such as the substitution of nuclear power plants for thermal plants, or vice versa. The interaction between our needs and the needs of other people results, in the open market, in the phenomenon of prices, which is discussed elsewhere in this book in more detail. Therefore, we can seize on the price structure as a good first step in setting up a useful value system. Without going too much into the obvious, the power of the price system to influence behavior affecting the stabilization of ecosystems is entirely credible; the power of pricing is acknowledged by all of us in our daily life. We all grumble, but we adjust to it. It may or may not be of consolation to us personally, but it is also a fact that under certain assumptions it can be shown that the price system often does organize human wants in the best possible way. One can well imagine that such a system might be a good basis for the regulation of troubled ecosystems.

Several difficulties come to mind. In the first place, the price-market system is best only under conditions of perfect competition. We must take this into account in accepting price as a good value system. Another trouble is that, even under perfect competition the price system may be the best way of satisfying existing wants for a given distribution of incomes, but it may not be fair in the sense that princes are richer than paupers and that resources that might be spent in growing potatoes are devoted to the production of caviar. This distribution effect may indeed be of some importance in ecological problems. But to the extent to which we all, princes and paupers alike, are bothered, and indeed threatened by air pollution, "income monopolies" are probably not our worst problem *in this context.* Society may well decide that social justice is a more important problem than pollution or that some mixture of priorities should be assigned, but to recognize this possibility does not imply we should let the two problems become confused. Anyway, this question must be examined in detail for specific cases.

So it seems that money, or the price system, can be used as a measure of value in at least some facets of the ecological situation. As a measure of value, then, it is correspondingly used as a criterion for the making of planning decisions for the modification of relevant ecological cycles and critically endangered ecosystems. But the major difficulty with using the price system as a method of evaluation or as a means of setting up planning criteria is that many benefits or evils connected with the ecological crisis appear to have no market value. The difficulties in extending the price system to the joys of a free-flowing, clear trout stream, for example, are formidable. To indicate this nonpriced nature of many human or esthetic factors, we may term the value units of esthetic and other quantitatively obscure factors as "supermoney." We will look into these questions in some more detail in Chapter 5.

Supermoney

We have emphasized that, if we try to use money (price) as the criterion for the selection of alternative courses in the fight against factors having undesired effects on the ecology, we immediately find that many important factors have no price at all in the usual sense of the word. Yet another example is clean air. I may want a smog-free environment, but nothing I can do will enable me to purchase it. It is, so to speak, as if ordinary money failed me, and I needed a type of super-money to buy what I wanted. In other terms, we can say there is no market for the exchange of money and clean air. Supermoney, then, can be defined as whatever actions would be necessary to buy ourselves a clean environment.

So supermoney is not a real medium of exchange but the mere ghost of a coinage that we could only wish really existed. Hopefully, we can eventually get the same results as we might with supermoney, if we intelligently reorganize the market with the assistance of government action.

In this book we will be concerned with methods of suggesting some kind of market for such things as clean air, so that clean air can be budgeted for in planning by government agencies just as trucks and mailmen are by the post office. In a nutshell, we want to transform nonproducts, or things having no market value, into regularly priced products. Naturally, comprehensive solutions to this type of problem probably lie in the far distant future. Even the derivation of some consensus among environmental experts as to the value judgment implied in evaluating the scenic value of the Grand Canyon, for example, is tremendously difficult. However, it is still exceedingly useful to explore approaches to such problems. Even a dabbling of an inquiring toe in the water, we think, is preferable to not getting wet at all. After all, any progress is good when the need is great, and the need is great here.

Chapters 4, 5, and 6, then, will treat, in order, ordinary economic criteria and some considerations of welfare economics; the role of what we call the human factor and qualitative criteria, the nonmarkets which we have just discussed; and finally, a special category of values that is discussed in an ecological context but that is too far from ordinary evaluation to treat in quantitative terms, or what we call "evolutionary mandates."

4

THE MARKETPLACE AND SOCIAL WELFARE: POSITIVE ECONOMIC CRITERIA FOR ECOSTABILIZATION CHOICES

Ecological decision makers, whether they are government officials, corporation planners, or housewives buying detergents, need what we may somewhat ponderously call "positive economic criteria" to aid them in choosing among alternatives. Once the features of the ecological and economic system are spelled out, the decision maker has to choose the best combination of those features. He may want a smog-free city and yet one in which automobile use is still free. If so, that is his criterion; but to enjoy the benefits of that world involves a cost, and this cost involves giving up other good things. Positive economic criteria will define the gains and the costs associated with possible courses of action for ecostabilization. Economic benefits and economic costs can then be measured in terms of these criteria. Then the following idea would appear to be a plausible concept, at least as the basis for a first cut at the problem: benefits should exceed costs in order for an ecostabilization program to be justified.

Because of the multi-faceted nature of ecological and economic systems and the positive correlation between the two, this chapter limits itself to the economist's concept of costs (not to be confused with the bookkeeper's) and the economist's concept of welfare (as opposed to soup lines). This material represents Thomas Carlyle's "dismal science" at its best (or worst?).

The Painful World of Choice

The world today is one in which there is no affluence, no abundance, and no plenty. Let's be hypothetical for a moment and consider a situation in which everyone in the world was told that he could have anything that he wanted, and in any quantity. Imagine, every individual in the world is told he may have all the goods and services he desires, *free,* and in any quantity. Imagine the production requests of the heads of the Pentagons throughout the world, of all the universities and other types of school systems, of the health services and police forces, and of the other multifarious public bodies and activities, if they could have all they asked for; imagine what business throughout the world would want and visualize the shopping lists of over three billion people. Now think (or if you like you may use a fourth generation computer) of what the sum total of the three lists—consumers, producers, and governmental bodies—would want. It is downright ridiculous to argue that that total would be something less than the material and human resources available in the world, but a very important point has been made. The nonavailability of resources results in the scarcity of goods and services for the world population, which means that wants or desires are not being satisfied.

Of course, the fundamental test of scarcity is price; however, in a world of vegetarians meat would not command a price, even if it were scarce; goods and services must also be useful. Usefulness and scarcity determine price. There are some goods, such as air, that are not scarce and, therefore, do not command a price; although a good argument can be made that air is terribly useful. Unequivocally, air will command a price if it becomes scarce sometime in the future.

There is, then, a menu of choices, and each society must decide what that menu will be composed of. In a fully employed society, one in which all resources are being put to some use, that society must consider the alternatives when it makes a decision to have more of one thing and less of another. The oft-quoted classic example, probably because of its dramatic value, is that of guns and butter. The society can either allocate all of its resources to the production of guns or to the production of butter. If the society is willing to give up some guns, it can have some butter, and vice versa. If we add to the menu unpolluted air, unpolluted rivers, etc., assuming fully employed resources, then the society will have to give up some guns, some butter, or some guns and butter. The factor of choice is the economic law of life. In choosing to add other items to our menu, we must give up some things, or a combination of things. However, if technology changes, it may be possible to add to our menu without the sacrifice. Given this event, the question is, of course, do we want to add more and more to our menu or even keep the same menu if it continues to add to the disturbance of our ecological system? Once again, a choice must be made.

When a society is at less than full employment, i.e., when there are idle resources, then we have a horse of a different color. With unemployed resources a society is producing something less than it could. By putting these resources to work, more guns and more butter could be produced. Although this may appear to be an exception to the economic law of life, it is not. In choosing or directing idle resource use it is true that neither guns nor butter is sacrificed, i.e., there may be a money cost but no social cost; however, choices will have to be made as to where these resources should be employed in order to achieve the society's best goals. These best goals may include full labor employment at the cost of ecoinstability, which is a social cost. Market guidance will be helpful in determining how to use these idle resources in the most beneficial way.

Now superimpose upon the individual choices of the members of this society the choices of the society as a whole: full employment, price stability, economic growth, and national security. The economic system truly becomes much more complex. One radical alternative to resolving the smog problem would be the abandonment of that civilized extension of the human leg (and national toy), the automobile. Visualize, if you will, the steel no longer required, the tires no longer required, the new and used car dealers no longer required, and the highway program no longer required, to name a few items, and the unemployment level that would result, not to mention the results of the next political election, if one materializes before the inevitable revolution!

At any rate, supposing we picked a less radical example, we can see that, in the short run, the resources at the disposal of the society would not be fully utilized. How these resources would be reallocated in the long run is an important question. But that reallocation can safely be left to the marketplace, if our planning for ecostabilization is intelligently carried out.

The Nature of Cost

Choice is an economic law, and there can be little doubt that it is a painful process, especially when pursuing ecostabilization programs in today's technologically complex and overpopulated societies. It is, therefore, quite important to consider and evaluate the *alternative,* or *opportunity* costs— the economists' concept of cost—associated with these programs.

Alternative, or *opportunity* cost (they are used synonymously), means that the cost of anything is the value of the alternative, or the opportunity that is sacrificed. More completely: *alternative cost* is the expected profit, or other return, like interest, forfeited when a particular opportunity is rejected in favor of another. This amount is then added to the cost of the other opportunity to make its total cost. To be sure, this is a broader notion about cost than the bookkeeper's, whose vision is limited to those costs that are actual cash payments. The account of society, the debit and credit records of economic viability, should go beyond that. The economist does. He realizes that the real costs attributed to doing one thing are not the money costs of that undertaking, but rather, costs stem from foregone opportunities that have been sacrificed. For example, the alternative cost of producing fuel oil in a refinery is the value of the gasoline that could have been produced from the same crude oil plus the expected profit that is forfeited. The cost of a vacation in Vietnam is the foregoing of a new automobile plus the interest not received because the money wasn't invested. The economic cost of collecting moon rocks is the other opportunities sacrificed. The alternative costs of having a baby are . . . , and so on. You can supply your own examples.

Discussion and evaluation of the national ecological problem must look beyond budgeted dollars and at the real cost involved in putting resources to this use. Thus, the use of the economist's concept of *alternative cost* is to compare benefits —those anticipated and those foregone. This, of course, is the familiar guns and butter problem again.

One meaningful measure of alternative costs for the United States, and other economies, is the technique of "input-output analysis," as first developed by Wassily Leontief.[1] The methodology employed reckons with intermediate sales and purchases (outputs and inputs) in the flow of goods and services from industrial sector to industrial sector to final consumption or purchase in the marketplace; it shows the interdependence of one sector on another. This interdependence arises out of the fact that each industry employs (generally speaking, of course, since some do not in practice) the output of other industries as its raw material. Its output, in turn, may be used as a productive factor by other industries. For example, steel is used in the production of trucks, and trucks, in turn, are used to transport the coal and pig iron used in its manufacture.

One of the basic uses of I/O (input-output) tables is to measure alternative costs. For example, assuming full employment of resources and a given state of technology, the foregone opportunities of a multi-billion dollar clean air program can be measured in terms of textile goods, lumber, leather, heating, plumbing, and so on. Of course, this gets complicated; the latest U.S. Commerce Department version on I/O tables covers 370 industries.

I/O analysis is used to show what can be left over for final consumption (personal consumption, expenditure, investment, government, and exports) and how much of each output is used up by various sectors in the course of their productive activities to produce some net output. I/O analysis has also been applied to the prediction of future production requirements, outlining marketing possibilities, the economic effects of armament, and more recently, disarmament, to name a few. This technique can also be applied to the ecostabilization problem.[2]

At the aggregated level I/O analysis may also be employed in assessing the relative "bad," (a term that takes on economic meaning in the section, "Smokestacks and Hard Cash") in the various sectors that contribute to and disturb the ecosystem. Other uses include: the establishment of priorities for resolving the ecological problem; the impact of the introduction of ecological technology on the economic and corresponding ecological system as a whole; and the measurement of the alternative or opportunity costs in any given ecostabilization program.

A Digression on Utils

Utils are not little people who wear bright colors and have fur on their feet; rather, *utils* may be thought of as measurable cardinal quantities like inches or minutes, that is, a measure to compare how much utility (satisfaction, bliss, "ophelimity," etc.) an individual derives from activities or products. Like our guns-butter problem, buying products involves a sacrifice or a cost. It also involves benefits. The things bought are expected to give benefits to satisfy the needs of the individual. Economists use the term *utility* to denote the estimate of the benefits. It is a term that reappears later in this book, and its concept is developed below.

Utility means want-satisfying power; it is some property common to all commodities, such as the utility of gasoline, a bridge, champagne, or an appendectomy. *Utility* resides in the mind of the consumer. It is subjective, not objective. A commodity does not have to be necessarily useful, i.e., the commodity may satisfy a frivolous desire or even be one that some would consider immoral. The concept is ethically neutral. Aren't we glad something in this world is!

Neoclassical utility is a descendant of the obsolete pain-and-pleasure psychology of the classical economists of the nineteenth century. It also has had associations with various social philosophies; ". . . the greatest happiness of the greatest numbers." Even though the concept of utility has been under attack for several decades,[3] its simplicity and common sense appeal cause it not just to survive, but to thrive.[4]

Economists are not amateur psychologists, and in admitting as much, the utility concept has lost all but the faintest vestiges of a psychological flavoring. We make no pretense at knowing why people buy things; we only make the modest assumption that people buy things because of utility.

There is a definite relation between utility and the quantity of a commodity purchased. As the quantity purchased increases, total utility increases. But there is one important fact in life and that is, the more units of any given thing an individual has, the less he wants another of the same thing. So that the gain or increment, i.e., *marginal utility,* will, after some point, begin to diminish. Additional units of a thing have less want-satisfying power. The marginal addition to total utility is zero when total utility reaches a maximum, which, in economic terms, means the individual has all of the commodity he desires. Negative marginal utility would mean economically that the individual has too many cats or drinks too many chocolate milk shakes, i.e., the addition of one more thing takes away from the total satisfaction. With unimportant exceptions, diminishing marginal utility, assuming given tastes, prevails for each and every good or service and is called the *law of diminishing marginal utility.*

The *law of diminishing marginal utility* also applies to money income. The size of the money income represents the utility of a dollar to the individual. The consumer's desire for different quantities of the commodity is represented by the diminishing marginal utility of the commodity to him. Then, the equilibrium of the consumer for the commodity will be at the point where the marginal utility of the commodity equals its price, or where the marginal utility of the dollar is equal to the satisfaction of having that last unit.

The *law of demand* rests, also, on the principle of diminishing marginal utility, taste, and income. That is, given the individual's income, his desires, and the price for a particular good or service, the consumer will purchase up to that quantity whose marginal utility is equal to the marginal utility of the dollars represented by the price, or equilibrium. Obviously, there is no equilibrium for an individual consumer if his marginal utility is increasing: nor would the consumer purchase any good or service whose marginal utility is less than its price, although some people do and then proceed to hate themselves for it.

83

The total economic value or revenue of the good or service (price×quantity), however, does not measure total *welfare,* a term developed more completely in Section 4.3. The total economic value of air, so far, is zero; but its contribution to welfare is certainly larger, as those who live in regions of frequent temperature inversions can testify.

This gap between total utility and total economic value is called *consumer's surplus,* a gap that occurs because the consumer receives more than he pays for. How is this possible? Remembering the fundamental law of diminishing returns, it is easy to see how this surplus comes about. The consumer, in buying each unit of a commodity, pays only what the last (marginal) unit is worth, but the earlier units are worth more to him, i.e., their marginal utility is greater than their price. He enjoys a surplus on the earlier units. For example, consider the surplus that is generated by an individual who purchases coffee. Suppose the price of coffee is $3 per pound, and he is just willing to purchase one pound, for a total expenditure of $3. Obviously, in giving up $3 for one pound of coffee, total utility must also equal $3. In this case, there is no consumer's surplus. However, if the price of coffee is $2, and we observe our same individual buying, not one pound or three pounds, but exactly two pounds of coffee for a total expenditure of $4, then his total utility must be $5, and his surplus $1. Remember that consumer surplus is the difference between what the individual pays for some given quantity and what he is willing to pay. Therefore, since the first pound of coffee is worth $3 to him, and the second is worth $2, (which, correspondingly, equals $5), the total utility of $5 minus the total expenditure of $4 results in a consumer surplus of $1.

By now the reader's heart is beating faster in anticipation of the application of the total utility, marginal utility, and consumer surplus concepts. But the application of utility analysis to the solution of ecological problems faces the obstacle of the validity of interpersonal comparisons of utility. Suppose Ramsay and Anderson have the same income, live in the same community, do the same kind of work, etc.; the problem posed by economists is: do they get the same satisfaction from spending $100, driving a car, playing golf, or going fishing? It is not an easy thing to peek into the minds of men and to compare their satisfactions from consumption.

It has been argued that if utilities cannot be compared, then they cannot be added. But this view may be too strict. To adhere to it is to ignore the seriousness of the ecological problem. It is impossible to think and act on this issue without assuming that people are pretty much alike when it comes down to their capacity to survive. People, for many purposes, are pretty much alike, but we also realize that they are not identical, and when interpersonal comparisons are made, we admit they will be rough—but maybe not so very rough when environmental conditions become pathological.

Utility is obviously a key concept in the environmental problem. The total utility of water is incalculably high; in fact, life without it is impossible. Yet, it has a low price. Its price, however, is not determined by its total utility. Price is determined by the relative marginal utility—the utility of one gallon, which is normally so low that people cheerfully waste billions of gallons a day. If for one reason or another water is priced higher than its marginal utility, the last unit will not be sold. The total utility of a super highway system, such as in Los Angeles, can also be high, but if there is no toll (large amounts of consumer surplus), the people use it as much as they please. The marginal utility of its use is effectively zero.

Concentrate on the marginal utility and not on total utility; therein lies the key. The more there is of a commodity, the less the relative satisfaction of having one more unit. Air, so far a free good despite its vast usefulness, will command a price when it becomes less abundant, and the last liter will demand its rightful (high?) price.

Consumer surplus aids the economist in making correct social decisions. To cite one example,[5] suppose there is a polluted lake that would cost the community $100,000 to revitalize. If the use of the lake is free to all, i.e., no dollar revenues from its use are imposed, then the sum of utility to its users will represent consumer surplus. For simplicity, assume there are 1,000 users who are alike in income and in their benefit from the lake's use. In addition, assume these 1,000 users represent the total tax base and are not voting other people's taxes. If every individual enjoys $100 (the per capita cost of revitalization) or more of consumer surplus from the lake, they should all vote to clean up the lake. On the other hand, if the consumer surplus is something less, it is uneconomical for them to tax themselves for this public project.

Equivocally and subjectively, society has felt that the disutility associated with disturbed ecosystems has been less than the subjective utility in consuming those goods and services produced over time. A societal equilibrium would be at that point where the equated marginal disutility of production is equal to the marginal utility of consumption—the balancing of these fundamental forces is an equilibrium point. Clearly, when the marginal utility of consumption is less than the disutility of production, cybernetic mechanisms should force the system back to equilibrium once again, unless, of course, there are constraints that prevent a positive adjustment, or no mechanism at all. This last case is all too common, alas!

Smokestacks and Hard Cash

We now take into account the costs associated with what economists call external *economies* and *diseconomies*, or *externalities*. These externalities are of concern to those economists who involve themselves with policy issues. More formally, they are part of the subject matter of a branch of economics called "welfare economics," which investigates the well-being of consumers and producers as persons and possible ways of improving that well-being through policy recommendations. This is what economists call "normative economics," (what ought to be) or the "art of economics," as opposed to "positive economics" (what is). Normative economics, however, is not independent of positive economics, i.e., economists, in making policy recommendations (to the displeasure of many and applause of a few), base their conclusions of doing one thing or another—implicitly or explicitly—on positive economics.[6] Economists are entitled, by their diplomas, to make such policy recommendations and are called experts. In the normative world these experts may or may not agree with one another.

Too often welfare economics is confused with the social service activities provided by the government, or "the welfare state." There is only a remote connection between the two. Economic welfare is confined to the subjective satisfactions of people as producers and consumers—satisfactions that can be evaluated by the measuring rods of money and supermoney (see Chapter 3). Welfare economics operates under the idealistic assumption that the government is omnipotent in exercising policies related to established economic welfare standards. Often enough these policies fall short of the ideal, yet they do have important effects.[7] This fact encourages the welfare economist to point to ways for bringing the real closer to the ideal—to lay out the path of economic welfare and the levels of satisfaction for a clean environment.

Assume that the economy is at full employment and resources are moved, by the market mechanism, from one industry or sector into another. To repeat our economic law, if we produce more guns there will be less farm labor to produce butter. More important, with respect to the ecostabilization problem, it may be highly desirable to increase the output of some product because it takes away from the pollution problem, even though it results in the sacrifice of a superficially more valuable butter product. We are now in an area in which the allocation of resources between two or more items is a matter of the relative urgency of the demands for them, their relative production cost, and their effect on man's environment. If the production of some good or service causes soil erosion, pollution of water, or contamination of the air, the decision maker must take these matters into account when deciding in what quantity the good or service should be produced or whether it should be produced at all. It is the difference between social and private costs and benefits and their consequences for optimal policy that will be discussed below and in Chapter 7.

External Diseconomies and Economies of Production and Consumption

The presence of external diseconomies means that social cost exceeds private cost, or that social benefits are less than private benefits. Social cost includes smoke producing agents, pollution of rivers, soil erosion, destruction of wildlife, etc. Private costs are those costs incurred by the firm (or industry) in its daily operations to produce some output and include the use of two or more of the factors of production, i.e., land, labor, capital, and entrepreneurship. The employment of these factors results in a certain cost curve of the firm, and in the long-run the firm must cover these costs or go out of business. However, all costs may not be included in the industry's cost curve; that is, there may be other costs besides those incurred by the firm. A. C. Pigou in his book, *Economics of Welfare*,[8] uses the example of smoke, where chimneys of factories in an industry belch forth chemical compounds that cause higher cleaning costs, higher medical costs, a loss of a view of the mountains, etc. for the people living in the same community. These costs are not borne by the industries that produce smoke in addition to their other output.

A modern day example [9] is the city of Steubenville, Ohio, a mill town (population 34,000) which the National Air Pollution Control Administration has tentatively ranked number one: ". . . the most polluted city in the United States." On a bad day residents claim that they cannot see the hills a half-mile away. Recently, a freak thunderstorm brought down quantities of hydrogen sulfide suspended in the sky and turned several hundred houses pitch black. The grass in the fields south of town turned iridescent blue due to other unknown pollutants, causing cattle to lose both their teeth and their appetites. On the average, 50½ tons of grime are dumped on every square mile of the Steubenville area every year. These other costs are not contained in the cost curves of the 125 or so industrial plants around the city.

The cost curves of the firms in industry exhibit the private costs. Social costs, to repeat, are pollution, noise, and other diseconomies caused, but not paid for by the producing agent. If firms within the industry had to bear the social costs, the cost curves of these firms would be higher. If costs rise, some firms may be forced out of the industry, or if they are monopolistic in nature, they may use prices to cover the additional costs. In both instances the supply curve shifts, resulting in smaller output and higher prices. Whenever social costs exceed private costs, the equilibrium output of any given industry (in the absence of government intervention or some other constraint) gives an output that is too large. Governmental intervention in the ecological arena may be viewed as a means of correcting this divergence.

How did these external diseconomies come about? Many are simply due to an expansion of the scale of a firm's operation. This requires additional inputs to produce additional outputs, and these increased inputs generate additional solid and liquid wastes. These wastes come to exceed the assimilative capacity of the environment, itself a decreasing natural resource.

Literature abounds with illustrations of this phenomenon. Expansion of operations could result in keeping, say, more airplanes in use, which is an added congestion factor to air use. Increased whaling efforts by one fleet of whalers depletes the supply of whales and makes it harder for others to obtain their catch. Increased use of water or more drilling of oil wells can make it harder for others to get these resources. Farming methods that erode the soil make it difficult for neighbors to produce and maintain the fertility of their territories. The reader can think of countless other externalities that make a strong case for supplementing complete individualism by some kind of group action.

Increases in consumption can also cause analogous disadvantages to others. For example, those individuals who purchase smog producing automobiles make life harder for others. Litter deposits take away others' enjoyment of national parks. Again, the reader can augment the listing of this type of externality; however, we add the fact that consumption patterns are heavily conditioned by society. The consumption of a new automobile or a baby is regarded as a milestone of economic and physical achievement; but these demands or consumption patterns affect the consumption patterns of others, creating diseconomics of consumption that are far from negligible. Sound economics would suggest some limitation on individual freedom in the interest of all.

We need to add that there are also external economies (as opposed to diseconomies) of production and consumption. For example, a firm, by expanding its operations, may be of a direct service to others, e.g., by the training of a labor force. Competitors will incur no training costs if they recruit any of these skilled workers. And when a firm expands its operations it may make it cheaper to supply services to all the firms within the industry. A textbook example often used shows a rise in the production of Ford automobiles that results in an increase in steel production. If there are still economies with large-scale production in steel manufacturing, the price of steel may fall. So Ford's competitors will also obtain this raw material more cheaply, simply because Ford increased the output of his cars. Again, there is a divergence between private and social returns, i.e., there is no remuneration to the firm training labor or to the expanding Ford firm for these benefits which it has conferred on others.

External economies of consumption can also cause analogous advantages which are not reflected in the returns of the person who produces them. For example, if Ramsay purchases more education for his children, he will make them better citizens. This confers an advantage on Anderson and others, i.e., it makes it possible for others to achieve a given level of satisfaction with a smaller total expenditure of their own resources.

Welfare Criteria

As mentioned earlier, welfare economics is concerned with the possible ways of improving the well-being of individuals through policy recommendations. Economists supply the decision maker with standards with which to appraise programs and to formulate policies. Although the economist is more at home in appraising the efficiency of the economic system, (certainly this efficiency provides a bench mark for appraising the system) it is not the sine qua non of society's interest. The immediate urgency of what ought to be done to disturbed ecosystems, and not efficiency reports, is of most concern to the citizenry. An accurate description of what is now, and the possible consequence of policy action provide that kind of contribution required by the individuals in this society.

Various criteria have been formulated by outstanding economists to distinguish between policy changes that improve welfare and those that worsen things. For example, Vilfredo Pareto (an Italian economist who had a major role in the development of welfare economics) developed the criterion that a change that harms no one and benefits someone must be considered an improvement. To say the least, this is a highly restrictive assumption. The elimination of automobiles, for example, would undoubtedly benefit many; but it would still injure the owners of the automobiles and also the manufacturers. So we cannot treat this interesting case by the Pareto criterion.

The Kaldor–Hicks (British economists) compensation principle was an attempt to broaden the applicability of welfare economics by adding to the Paretian optimum case those situations in which individuals benefited would be willing to compensate those who lost to the extent that the latter would no longer be opposed to any change. To return to our original example, those individuals enjoying blue skies and clean air would be willing to pay automobile drivers for not driving. An interesting twist is that they may be the same people. One might well think that the problem of ascertaining the compensation to be offered and accepted would be an economist's nightmare, in general. But if they were indeed the same people, that would eliminate the need of gross compensation transfers. Only net transfers would have to be made, thereby cutting the range of error, and if the compensations were generally equal for all the people, the need for any transfer at all would be eliminated. Instead of a nightmare, it would be, and it is a source of simplification for which we may be grateful.

Various social welfare function criteria considered by the economists Arrow, Samuelson, and Bergson are commonly studied today. These functions express and weigh benefits and injuries in various manners, in terms of the nature of the gain or loss—cleaner air in preference to automobile use, for example. It is true that attempts to construct social welfare functions often run into trouble, e.g., the Arrow "impossibility theorem." [10] So further theoretical research is needed in examining the basic requirements of social welfare. In the meantime, the world does go on, and it is not impossible to develop plausible, if imperfect welfare functions for major segments of our economy, and to make judgments rather reasonably about the desirability or undesirability of certain policies.

A noneconomist can argue that restrictions on the firms pouring out noxious fumes in Steubenville, Ohio can be accepted on the basis of a social welfare function that dictates the benefits to those individuals living in the area are greater than the injury to the firms. He would argue that the desires of the community are obvious. But the desires of the community are not often obvious to the community members themselves, and where they are, they can be on the side of more production and more jobs, and hang the pollution. Keep in mind that the main job at hand for the economist is to identify the real social and individual costs and the real social and individual benefits, educate the public about them, and advise the public of possible solutions.

5

**OUTSIDE THE
MARKETPLACE:
EXTERNALITIES AND
THE HUMAN FACTOR**

We know quite a bit about the economic life of man, or the way he behaves in the marketplace. While the science of economics is as full of uncertainties and controversial theories as any of the other fields of social study, economics does have a common unit of value and well established ways of equating dissimilar quantities. And the necessity for economic analysis is accepted by those who are concerned with the practical building of the world. For example, the phrase, "engineering economics," has a plausibility that might not be accorded to, for example, "engineering sociology." Still, the fact remains that the whole of life is not contained within the marketplace. Yet the problems of ecology touch on many of these noneconomic parts of life. If we loosely denote the set of these noneconomic, or nonmarketable facts of life as the "human factor," then we see that we must examine the needs of human beings as expressed in noneconomic actions in order to be able to make sensible planning models for ecostabilization efforts. Since it is usually difficult to measure these human factors entering as criteria into the planning process, we may also call them "qualitative criteria." Our purpose here is to try to get these qualitative criteria into as quantitative a form as possible and, then, to connect them with the externalities of production that are of concern to us in environmental problems.

The Problems of Being Human

The last two centuries have seen great spurts of progress in the scientific investigation of why people act as they do. This progress in the social sciences seems pitifully small, compared to the rather spectacular technological achievements that have derived from the growth of knowledge in the physical sciences. This notorious lag between the social and physical sciences is unfortunate for science, in general, in that the subject matter of one particular branch of science often depends on the state of knowledge in another branch. When one of the two branches is poorly developed, it may be very hard to make progress in either science.

Such a bottleneck was present in the physical sciences until the 1920's. The periodic table of elements was a kind of mysterious fluke of nature. The table depicted regularities in chemical behavior of substances, but the table itself was just a fact of life. The facts of chemistry concerning atomic structure did not become understandable until the quantum theory had first been developed in physics. Once scientists understood that electrons had some qualities that were more like those of light waves than those of Ping-Pong balls, the almost magical nature of the periodic table of chemical elements became explicable at a deeper level of understanding.

Analogies within the social sciences are many, but we can see that a relating of economics to psychology is a particularly salient example. After all, economics depends on the approximation that man behaves rationally as a *homo oeconomicus,* or economic man, and, therefore, prices and markets

95

are abstract things that represent customary behavior by men. Obviously, this whole concept goes up in smoke if people suddenly decide that they would rather give things away than sell them, or that all consumer goods of a certain kind should be destroyed for irrational reasons, as in the case of Savonarola's revolution in Florence in the fifteenth century, with the public burning of objects of vanity. And it is in fact quite easy to make economic man disappear at least temporarily. A war or a riot can destroy the marketing and production system of an area and make it appear as if they had never existed at all. Also, the subsequent growth of barter markets, or "cigarette money," after political upheavals can be considered as a recrudescence of man the economic animal.

As in this book, when we try to suggest ways of extending the types of analysis used in economics to the externalities of production and of other factors not normally seen in economic markets, we are naturally concerned with the neglect of rational and irrational factors in the economic approximation to the nature of man. It is true that rational factors that are omitted from the economic view can be included in a generalized formulation almost by definition. After all, the word rational implies that laws of cause and effect are understood and can, therefore, be expressed, at least in principle, in the no-nonsense logic of mathematics. The so-called irrational factors are another story. What is strictly irrational cannot be included in planning models in a sensible way. We must realize the labeling as irrational of a human act such as a murder, a dream, or a thrill of passion, tells us not something about the basic nature of the act, but only about our understanding of it. In this regard, the history of psychoanalysis provides a cogent example. Freud and other psychoanalysts, regardless of the controversial scientific foundation of many of their assertions, have been responsible for a veritable intellectual revolution in our understanding of the behavior of human thought and desires. Their influence has been responsible for an appreciation of how important irrational thoughts are to all of us. In that sense, psychoanalysis and related clinical psychological methods would appear to have a discouraging effect on planning methods which attempt to satisfy the rational needs of man. But it must be noted that psychoanalytic reasoning itself is a rational process. Understanding what has been termed the irrational element in man has helped to make many of our hidden desires and fears conducive to logical analysis. The importance of the subconscious in determining the human experience has, it is true, been established. At the same time, we have gained a rational understanding of how that subconscious shapes our destinies.

So the problem of placing the total human being within the perspective of scientific planning, although it faces tremendous difficulties, nevertheless affords us hope of future progress. Besides, many problems that are tremendously difficult in theory do seem to work out in practice, if only by the laws of statistical averaging. A good example is the well-known human need for variety in activity. One of the most critical hidden problems of human economic existence is yet unsolved: the fact that it pays society to encourage workers to specialize in certain tasks, while the natural human desire of every person is to experience variety in his activities. This need for variety also shows up in the changes in women's fashions (and currently, in men's haircuts).[1] Since these changes often require additional spending by the individual, they are subject to continual consumer criticism. But certainly we can look on styles and fashions as a saving grace, as a patently human quality. Also, we must notice that even though planning for styles is either impossible, or a carefully guarded secret of the most successful couturiers, the price-market system, a rational entity, has no difficulty at all in handling the sale of clothing fashions.

Much of our reluctance to extend the realm of the rational

undoubtedly stems from motivations that are themselves hidden. We will return to this point in the next section.

Homo Nonoeconomicus

In looking at man in his nonoeconomic roles, a good place to start is with the extensions of simple value theory inside market economics itself. As we discussed previously, welfare economics, or what we can define in a more general form as *metaeconomics* or the extensions of economic theory beyond the market place, attempts to deal with the satisfaction of those human wants and values that may be neglected in the usual market economy. In the market economy, of course, we can measure values by money, the medium of exchange for equating utilities of various goods and services. If we go beyond the market, how can this utility be measured? One answer is to use people's preferences in terms of votes to determine production choices for the society. This approach requires that in some way the preferences of the individual must be converted into the preferences of society. Unfortunately, the "impossibility theorem" mentioned in Chapter 4, which seems to make difficult the converting of individual preference orderings into societal preference orderings, has been the subject of much frustrated study in recent years.[2] Furthermore, there seems to be legitimate doubt that the welfare of society depends strictly on individual preferences. The hooker here seems to be in the word "prefer." A man might choose to be faithful to his wife when tempted by his secretary, but that does not necessarily mean that he wants his wife more. He has a preference based on moral reasons or inhibitions, a preference that might be described by "would choose for any reason."[3] But this does not mean that he would vote for fewer secretaries and more wives if we consulted his preferences on a pure, or "wants more" basis. Given these difficulties with the "analyzing everything by individual preference method," there seems to be no reason why the utilitarian social welfare function,[4] with its assumption of common value measures, cannot be used as the basis of a planning function for ecological purposes. The utilitarian assumption that welfare functions should be optimized to produce the greatest utility per capita is at least free from the difficulties of the impossibility theorem.[5]

The utilitarian, or "best on the average" criterion, of course, does us no good at all on questions of social justice. A society of millionaires and paupers is a possible utilitarian optimum. But from a purely political or practical point of view, certain considerations of equity can be quite easily included in the overall welfare calculation.

If we accept, at least heuristically, the utilitarian point of view in regard to welfare economics, we can start out with a criterion function free of obvious internal logical difficulties. If we take the further big step of equating utility with market expressions of utility, the social welfare function can then be equated to the measurement of net output per capita in a society, with one very difficult proviso. All economic externalities and noneconomic factors of any importance must be valued in ways analogous to market pricing. Can such a difficult task be accomplished to any useful degree of approximation?

The economist Kenneth Boulding has recently pointed out some difficulties in generalizing the valuation coefficients (in other words, "prices") of economics to other problem areas.[6] He believes that even after all the neighborhood effects, external economies and diseconomies have been dealt with (and the "seven marginal conditions" of welfare economics have been fulfilled),[7] there remains a residue of elements in social relations that are not accounted for by exchange.[8] He calls these residues an "integrative system" and considers this system to involve things such as status, respect, love, honor, loyalty, identity, and legitimacy. We can readily believe at least part of this argument if we realize that the marketplace itself is founded on such factors as the trust and credibility of buyer and seller.

At this time we think that no final negative argument in this difficult area can credibly be made. Even Boulding asserts that a strong case can be made for the extension of at least the method of economics into the integrative system, so that all human values become at least subject to a quasi-economic analysis.[9] In addition, it has been held that many values in the integrated system do not represent absolute requirements, or, in economic terms, infinite prices.[10] So there is no reason why one may not try to fit a normative calculus like cost benefit analysis (looking for the biggest "social profit," or in the jargon of the Department of Defense, "biggest bang for the buck") to various noneconomic values. By the use of what might playfully be named a "descriptive-positive" approach, the function of existing mechanisms that set monetary values on human factors can be studied in order to set prices for noneconomic goods. A suitably cautious version of this approach is what we suggest here.

We find a somewhat different difficulty in extending the tools of economic analysis to the consideration of human values—the possible neglect of ethical considerations.[11] Of course, to the extent to which the ecological decision problems are mostly strategic and not ethical, the question does not arise in any important way. Some sort of ethics certainly lies behind all of the preferences and values expressed in any system. But we can take the handy conservative viewpoint that the ethical problems cannot be decided by utilitarian analysis, but must be evaluated beforehand. In other words, the ethics of the problem enter into only the setting of a price on quasieconomic human values, not in the balancing of prices and the costs which go into the overall welfare optimization. Using an example of twelve men in an overcrowded life boat,[12] it might be supposed that a naïve utilitarian calculation would justify throwing overboard the old and the sick first. Another, possibly more ethical calculation might recommend choosing those to be sacrificed by drawing straws. It seems evident, however, that such choices do not involve the method of optimization of a welfare function, but involve the pricing of human life itself. The "naïve" utilitarian approach might assume that human lives are valued differently, depending on physical condition, while the other approach supposes a constant value for each human existence. Such a decision is surely a pricing decision and should come before an optimization calculation rather than be the result of one. Of course, no manipulation of the mathematical formalism can make such difficult problems go away, but a realization of their proper place in the formalism can prevent the generation of spurious results by a planning model.

Fortunately, the problem of assigning an appropriate monetary value or other *numéraire* (measurement unit) to human values is perhaps not as desperate as might appear at first. That is, the natural confusion in attempting valuations of human factors leads often to a misunderstanding of how they occur in an actual decision problem.[13] Often strategic questions, such as how to arrange certain factors to attain a given end, are the actual subject of disagreement. But since many strategic problems, such as the monetary and fiscal policies of the United States government, are extremely complex, disagreements on strategies are often attributed, instead, to disagreements on end-values or goals. Here again, the credible conversion of qualitative criteria into quantitative terms requires a careful sorting out of the sheep from the goats.

At any rate, now that we have chosen the point of view to be followed, that of an extended cost-benefit analysis applied to economic and noneconomic, or "human" values, we can examine some of the directions that may be taken in making these evaluations for features of the ecological problem.

Converting Damns into Dollars

Tremendous possibilities open up for us if we can hope to treat, in some kind of sensible approximation, the "integrative system" of human values mentioned above in the framework of some sort of social accounting system. Specifically, the severe problems of the ecological crisis can then be treated by mathematical means. This does not mean that we use the mathematics to do our thinking for us; it only means that the mathematical framework, or model, can be used to keep our accounting straight. It is very much like the accounting of a business firm. A salesman fills out his expense account according to different categories, such as meals, lodging, transportation, and phone calls. These expenses have definite places in the ledgers of the company. In turn, these ledgers will be used to determine what the profit or loss of the firm is, and later they will be used in an analysis to find out how the operations of the company should be changed. This does not mean that the expense account cannot be questioned. The expense account is constructed by a finite being, the salesman, having finite amounts of intelligence and honesty. It must always be subject to audit. In the same way, planning for gross areas of human existence must always require constant audit of the difficult socio-economic decisions that lie in back of planning models. With this in mind, we can examine briefly how the prices of "priceless" quantities, such as clean air, can be established. And we can see how these prices can be used by government and private planners to adjust the markets of the world in order to help alleviate the problems of ecological imbalance or environmental disruption. Some more specific examples of the establishment of prices, the treatment of taxes and compensation, and the assignment of model parameters for externalities and other noneconomic factors are treated in Appendix B and Appendix C.

The Uses of Ignoble Behavior

It is said that every man has his price. That price may be money; it may be power; it may be fame or love. In cases of determined narcissism, the price may be beyond what any reasonable person would pay. Assume that we wanted to buy one or more men for some social purpose, such as building a pyramid, and that all the men were more or less equivalent for our purposes, that of dragging stones up from the Nile (or the Hudson) to the bluffs above. If we wanted to get as much for our commodities as possible, we would first of all have to know how much money was worth, how much power was worth, how much love, and so on. This, of course, is the crux of our problem, that of establishing an overall valuation. For planning purposes, an overall valuation must be defined, even though the precise form of the common measure (or *numéraire*) is undetermined. Can money perform service as a *numéraire?* Is it not obvious that it can, and that only an unwillingness to admit to the so-called ignoble side of man's nature stands in the way of this admission? Each of us must supply our own answer as to whether money can buy, for example, respect, fame, and love.

For scientific work, however, a yes-or-no answer is not enough. The scientist can not be satisfied by such general assertions about human values. Also, to the social scientist the consistency of valuation is as important as the amount of it. Naturally, we are still a long way from a rigorous treatment of this consistency problem, but the crisis of our environment, as well as ordinary curiosity, encourages us to take some tentative steps in this direction anyway. So damning the torpedoes, we try to look at the problem of evaluating human values as they appear in the externalities or external diseconomies of the ecoproblem.

A good example of the type of externality which is difficult to measure is the negative value of polluted air in the Los

103

Angeles Basin. It is certainly true that each of the inhabitants attacked by the ozone and nitrogen dioxide of a particularly bad smoggy day would be glad to contribute a small sum to make the pollution go away. Even such a sum as twenty cents from each of the inhabitants of the Basin would insure that a million dollars could be contributed on any one day toward the cost of removing the pollution. Unfortunately, it is also true that the adaptation of motor vehicles to a less pollutant state would require much greater sums than that, even if the informal collection could actually be carried out. So it becomes very important to find out exactly how much the smoggy air is not worth to the people of the Basin. How important to us is the possibility that the physical discomfort of today's attack will actually produce diseases, such as emphysema, lung cancer or heart disease, tomorrow? Also, how about the unperceived and unevaluated smog components? Certainly the presence of carbon monoxide causes little noticeable discomfort; yet recent studies have linked it to greater probability of cardiac malfunction.

Incidentally, we can see that there is a knotty question right at the start. There are two possible smog (negative) values that a person might consider. One is the amount that a person would be willing to pay in order to avoid the air pollution on the basis of the facts he knows. The other is what he should pay on the basis of all information available. We propose here to use the stabilizing mechanisms of the price market system to effect social ends. So we suggest looking for values of perceived wants, just as the ordinary shopper in the market does. There is obviously no law against the use of education and public relations techniques to heighten the value of the wants that are experienced, but in any kind of political and economic democracy, no matter how imperfect, true ecostabilization can only take place on the basis of a generally agreed upon method of evaluation. This means what people do think, not what they should think.

This is all very well in theory. But how can we evaluate smog in practice? It is consistent with the price-market approach to follow a method that might be called "deterministic pragmatism." An example of evaluating smog by this method would be to look at the motives of people who vote on the issue with their feet, that is, people who leave the Los Angeles Basin with the avowed purpose of escaping the smog. Naturally, the human tendency to add excess drama to other decision variables must be reckoned with. Leaving an area because of smog is a very high-prestige excuse, covering anything from a job layoff to rubber checks; but a study of such *émigrés* would still serve to help determine how much the negative value of smog is. The Los Angeles Basin is, in general, a region of high wages. After all extraneous factors have been averaged out, such as the days of sunshine, the relative absence of rain, the lack of distinctive seasons, and the possible presence or absence of mothers or mothers-in-law, a value set by emigrating people in terms of lost-wage opportunities might give an indication of the actual market value of smog. Certainly, a study of other areas in the country would serve to make possible a more sensitive correlation, or regression analysis of the data.

No one has carried out such studies. Reliability problems would probably be immense. Other, perhaps more practical tacks have been taken. For example, preferences for living area are expressed in the marketplace by real estate prices, and the real estate market is very competitive. So the investigation of land values, comparing smog-free to smoggy environments, seems a promising approach.

The effect of pollution on decreasing urban land values has actually been the subject of detailed studies.[14] Studying the situation in St. Louis, Ronald G. Ridker tried to find out how property values tied into the amount of air pollution. The air pollution level (as measured by "sulfation levels," or the concentration of sulfur trioxide) was determined for a large number of census tracts in the city. These pollution levels were then compared to property values. Of course, the question of what property to compare to which is complex; about ten other factors were taken into account, so that the air pollution effect could be singled out: number of rooms per unit, density of housing, proportion of new units, travel time to centers, occupation and incomes of owners, etc. The results indicated, within stringent statistical limits,[15] that a reduction in the "sulfation level" to a chosen low level would increase single-family property values in St. Louis by at least $85 million. At 8 percent interest, this means that pollution was costing the city a minimum of about $7 million per year.

We should emphasize that even though the exact relation between property values and air pollution was not known beforehand, statistical techniques (see Chapter 8) made it possible here to carry out a scientific study of the smog valuation problem. One study, or several studies, has not and cannot solve the economic valuation problem. But it is important, not that we now know the exact answer to the problem of costing pollutants, but that progress can be and is being made.

Turning to water problems, we might ask, "How much is unpolluted water worth?" In areas of great water scarcity, polluted water can be rated as worth zero dollars, and the negative value of pollution can be treated as an opportunity cost for the use of fresh water. That is, water we have spoiled cannot be drunk, and we will have to buy an equivalent amount of fresh water to pay for it. That extra amount which we pay is the cost to us of the polluted water. In places like the offshore regions of southern California and in Lake Erie, where fishing industries have been crippled or destroyed, the value of the fishing output forms one direct measure of the pollution costs. The case for recreational waters is somewhat more difficult. But we must remember that the recreation industry is now quite organized, and statistics exist on the use of national and state parks, while figures are available on the issuance of fishing licenses and the sales of sporting equipment. Naturally, there will be discrepancies in a possible valuation of recreation established under such methods. But some idea can certainly be formed concerning the size of the values involved. Also, with the present rate of expansion of world population, it should certainly be desirable to place at least conservative limits on the actual money values of unpolluted water and air by the application of present average values, approximate though they may be.

At any rate, efforts in the economic analysis of recreation areas have actually been made. The question of the economics of recreation has been looked into from the point of view of market analysis parameters, and the theoretical dependence of recreational demand on transportation cost has been examined.[16] Also, another line of attack has concentrated on observing correlations between social facts (race, income, etc.) and the demand for swimming, boating, and fishing.[17] Two of these studies are summarized a little more completely in Appendix B. Difficulties in evaluating the benefits in terms of money still tend to crop up, of course. Concrete results are often limited to such measurements as recreation "activity days" gained or lost. But there seems to us to be a good argument for valuing days spent in recreation, e.g., at prices something like an average daily wage. At least the fact that standard overtime and Sunday wage rates are time-and-a-half or double-time can be used to suggest that the disutility of working corresponds somehow to the utility of recreation.

Other pollution problems may be more complex in their effects. Where natural standards, such as total disposal, are available, cost considerations alone can be used. For example, in the disposal of solid wastes, typical values of the costs of the total disposal of wastes in Southwestern Los Angeles County, from garbage can to sanitary landfill, are in excess of ten dollars per ton. For comparison purposes, this price exceeds twice the amount paid locally for good quality river sand for construction purposes. Waste is an expensive commodity! Of course, solid waste in the form of beer cans discarded along the road enters into the metaeconomic equation not only in disposal costs, but also in the form of human values, such as esthetic (negative) considerations. Even neglecting these difficult costs, there is also the solid economic disposal cost to use for conservative planning calculations, even though waste is not generally considered a commodity in the marketplace. It is ironic that the solid waste problem is often more straightforward than the gaseous waste (air pollution) problem. We do not have to balance benefits for and against garbage pollution; all we have to do is decide among various disposal methods (of course, in practice, some methods are relatively ineffective, in which case the "benefit decision" problem arises again). Sanitary landfills, which ruin scenic areas, certainly have negative benefits but so does incineration (double-chambered, hopefully), which contributes to air pollution. So difficult problems can still arise. But popular standards seem high, fortunately, in the field of solid waste disposal, so that planning steps are relatively easy to implement politically. People seem to have a somewhat lower tolerance for garbage on the ground than for garbage in the air.

Other examples could be quoted, but these few instances should give an idea of the kinds of problems and kinds of solutions we have to deal with in trying to find out just what economic values are involved in common pollution problems. The whole pollution cost problem, of course, also includes clean up costs per se, as opposed to damage costs. We do not want to go into the great practical difficulties of choice of clean up methods, methods of financing, etc. Such costs can usually be evaluated by observing market phenomena, while the damage costs or diseconomies often cannot. So the damages deserve that extra theoretical attention we give them here.

We have left until last the most frightening damage to our environment, that to our own life and health. To handle this problem in economic terms, we must be able to place a plausible price on health itself. Now, as we have mentioned, the harmful effect of air pollution, for example, on one's own health may be inadequately "priced" by the ordinary consumer because he lacks accurate information on harmful externalities. Nevertheless, we may be able to supplement such inadequacies of pricing from indirect sources. For example, if we consider that health care forms an important category of governmental expenditures, one can see that governmental planners can identify at least some health costs in purely money terms. From the point of view of the budget of a State Department of Health, the effect of smog is, in principle, a directly measurable quantity. Projections of per capita health cost in cities of varying degrees of air pollution, properly correlated for differences in income, population and other quantities, can lead to the setting of a price per unit of pollution in terms of tax revenues expended for health. From the point of view of producers, also, the effect of smog on absenteeism and lost labor hours due to sickness forms a direct cost to their operations. All such costs must be included in a proper planning framework.

A significant amount of research has already been carried out on the economic aspects of health. The cold-blooded effects of premature death and sickness on the labor supply in the United States have been estimated, by calculation of the surplus product, that workers, on the average, provide for other consumers.[18] An idea of the magnitude of the costs involved is given by the figure of $65 billion in extra production brought about during 1960 by mortality rate decreases since 1940.[19] A study of mortality, morbidity (sickness causing work time loss), and debility (sickness reducing productivity during working hours) in Puerto Rico found that the Commonwealth Water System costing a total of $81 million ("discounted," see Chapter 8) by the year 2004 will produce labor-health benefits of $172 million (also "discounted") [20] (see Appendix B for a brief resumé of this study).

The cost of health loss due to air pollution has also been examined.[21] Dirty air probably causes premature deaths and morbidity from respiratory diseases, such as cancer, bronchitis, colds, asthma, pneumonia and emphysema. Total cost for these diseases (including premature burial expenditures and costs of treatment) was calculated at about $2 billion for 1958. Other costs, such as removal of patients to healthier areas, would raise the total even higher. Of the total cost, about 20 percent was attributed to air pollution, or $400 million. The figures, though, while interesting (or alarming), are not the main point here. The point is that plausible, if not perfect calculations can be made. Death rates, treatment costs, and absenteeism records are all ordinary kinds of data that are relatively readily available. Even the proportion of illness due to air pollution can often be estimated credibly, in this case by comparing urban and rural incidence of the diseases studied.

A related question is the value of human life itself. Lost lives do correspond to lost production, but a human life is also a "consumer good," not just a cog in the social machine. As has been increasingly pointed out in research studies,[22] evidence of such human life values can be traced through various indirect means. Certainly, decisions on building roads to maximize highway safety can give clues for the valuation of human life. Again, mundane and unromantic as such considerations may be, life is subjected to all sorts of monetary evaluations, such as insurance compensation cases and various other court actions. The legal evaluations are, of course, notoriously variable. Still, hypotheses about various types of "regression equation" relations between loss of life and compensation payments can be made, and we can try to take the value of life itself into account in planning consideration in much the same way as health. Certainly, philosophers may argue justifiably against such materialistic methods, as the use of compensatory damage suits to place a value on life. But when the survival of mankind is at stake, a healthy dose of pragmatism seems to be called for. It seems appropriate that one of our most remarkable worldly institutions, the common law, may help us to keep the earth a place we can live in.

We should like, then, to encourage the treatment of externalities involving human factors so that they are then amenable to the direct accounting of economic methods and, consequently, to audit by democratic processes. Some more difficult problems in the esthetic sector of human factors will be mentioned briefly later, but now we have to consider how these prices of human factors can be used to help human beings plan their survival.

Getting Even

We can place prices of some sort on pollution and other ecological instabilities, but how can we develop a market for them? One obvious answer appears to be to use taxes to replace the usual market exchange of values. There is no market for clean air because we cannot sell our lack of automobile exhaust to our neighbors nor can the power company sell us a lack of carbon monoxide and nitrogen oxide from its smokestacks. But by acting through our government, we can tax ourselves and our industries as producers of smog. We can think of these taxes as acting to discourage pollution directly. But from an economic view, we can also think of them more generally as affording us a means of compensating ourselves for the environmental degradation, for compensation is indeed possible. There are some sacrifices we will not make for the environment. That means, in economic terms, that pollution has a price and that this price could be the basis for compensation for damages. It is important not to overstate the case against pollution, regardless of current rhetoric. Air pollution is a critical problem, but it is safe to say that clean air would not feed and house the hungry and homeless.

In the future, even in the near future, pollution may command an infinite price (meaning, it must be avoided at any cost), and this possibility must eventually be taken into account. But at present, we can consider compensating ourselves for our sufferings as at least a minimum environmental program. This compensation may take direct forms, such as the reduction of sales taxes to enable us to buy more consumer goods for the same amount of income, or we may decide on more complex solutions, such as using smog payments to establish rapid transit facilities. In either case, the general applicability of the compensation principle remains the same, although there may be legitimate differences of opinion about exact actions. What is important here is to establish the practicability of rational methods for controlling our destiny, methods on which reasonable men can agree, or which at least form a solid basis for political argument and democratic decision.

Legitimate questions arise about the practicability of tax compensations for taking care of damages from economic externalities. Some considerations in the theory of such schemes are discussed in Appendix C. We should say here again that taxing and compensation alone may not be enough, and that regulation may play an important role. If the pollution of our waters, for example, comes to threaten literally our total existence, we must reemphasize that the taxes necessary would become infinite; i.e., all relevant production processes would have to be stopped by legal prohibition.

Nothing we have said here should be construed to contradict this possibility. We only wish to point out that unwarranted use of fear of the law to gain ends can have all the features of social overkill. Too much may be done in too destructive a way. Under heavy-handed regulation (like Prohibition) it typically happens that the criminal prospers and the honest man suffers. Sophistication in the guidance of human affairs has been a hard-won ability. Let us use it to help us in the ecological crisis.

The Money Value of a
Pure and Perfect Environment

Ultimately, we hope to be able to apply a quasi-economic analysis to the nonmarket quantities involved in the ecological problem, whether they are in the form of the usual kinds of externalities or whether they are human values that lie outside the ordinary economic accounting entirely. We have seen how some of these factors can be managed. We are far from being able to reduce all qualitative factors to quantitative terms.

One glaring omission is the role of esthetics in the meta-economic scheme of things. All of us want a more beautiful and pleasing world, but when it gets down to a decision, how much are we willing to pay for it? Some work has been done on this problem in connection with the evaluation of wild rivers. We have seen how the negative value of pollution in rivers may be investigated by comparing expenditures on recreation. Of course, the lack of pollution in a wild stream is not the only factor contributing to its recreational value. Also, of course, recreational and esthetic values may conflict. We can dam up the wild river, creating more recreation in boating and swimming (at least by some measures), but we may also destroy the value of the valley in the view of the Sierra Club. Similarly, putting a highway through a scenic mountain area would make it more accessible to tourists. Each new tourist would, from an economic point of view, gain net recreational value. But we all know intuitively that the area would be spoiled. Marine discusses this type of conflict in his book, *America the Raped*.[23] He concludes that there must be a balance between areas easily accessible to the public and those which are kept as wilderness areas in the strictest sense of the term. In theory, all this could be handled from an economic point of view. In some way, which we cannot precisely put our finger on, the wild areas have a value to us, and this value should be, ideally, expressible in money terms, just like anything else. This value, if we knew it, would then be greater or lesser than the recreational value to the tourists who would come into the area through the proposed new highway. By comparing these two values, the highway department and the park department could make a decision for or against the road.

This theoretical picture is still distant from realization. But we can see that the conception is not inane, in that the relative values three hundred years ago (in America) and three hundred years, say, in the future should be obvious. Miles Standish would certainly have built the road. Miles X97324911 of the year 2270 A.D. would certainly not. Scarcity rules in economics: in one case roads were scarce, and in the future wild areas will be scarce (some people think they are scarce already).

So the problem is one of quantifying vague values today, so that we can make decisions intelligently. We think it is encouraging that at least some attempts have been made in this direction. Recent studies have suggested various schemes for setting relative values of wild rivers. One scheme [24] treats three "more important" factors: the quality and appearance of the water in the stream, the vegetation of the flood plain and surrounding valley, and the aquatic and terrestrial habitat for various species. Four "less important" factors in this same suggested method are "the view of the valley from above, the view of the valley from below, the flood plain vista, and the appearance of the channel." Various weights are assigned, rather arbitrarily, to the different factors. Another scheme quantifies the landscape esthetics in rivers and valleys, as well as in terms of fourteen "physical factors," fourteen "biological and water quality factors," and eighteen "human use and interest factors." [25] The latter includes the degree of urbanization, artificial control of flow, accessibility, historical interest, and special use of landscape features. This is certainly a formidable list!

It does not appear that such schemes are very promising in their present form, although some efforts have been made to use them in cost-benefit calculations.[26] The problem is difficult, and research efforts in this direction should be encouraged. Intuitively, we know that these human or esthetic values, beyond purely recreational values, have some monetary equivalent, since we are willing to spend money to enjoy them. In estimating the difficulty of the task, we must realize that esthetic values do not have to be the same for all of us. The price system has always functioned by means of the "decomposability" or "factorability" property of individualistic consumption.[27] The river need not have the same utility to a fisherman and a nonfisherman, in order for us to set a price on it, any more than a five dollar steak dinner has equal utility to a hungry football player and to Miss Gibbs, the vegetarian church organist.

The esthetic question is, of course, only a particular facet of the general hazy area of human values, but it is an important one for ecological purposes. The general problem of including all human activities into one valuation system has been the subject of sociological investigation. As an example, one suggested plan of attack [28] considers an economic market, a governmental structure, and "organizational markets." In these "organizational markets," various human values are treated at least in part as commodities. These commodities are denoted by the prefix "c," so that one considers the commodities c-respect, c-affection, c-rectitude, c-well-being, c-enlightenment. These commodities can then be traded in the organizational market. The importance of considering these human values in economic planning is pointed out: after all, cultural values do restrict the operation of the economy in ways not ordinarily described in economic input-output models. In addition, political groups seek to maximize c-power and may perturb economic planning (as discussed in connection with some examples in Appendix C, the problem of political values can be included, at least, on an ad hoc basis, in planning equations). In addition, c-power is often a precondition for any planning at all, so that the satisfaction of such constraints is a sine qua non of any further actions. But the fact that people do have desires for the c-human values is unmistakable. Unfortunately, this model (Isard-Rydell) is understandably not yet able to express the human c-factors in the same value system as economic factors. If we could only treat c-esthetics and c-power in some quantitative way, what a help that would be for making environmental decisions!

We have seen how indirect evidence on human values may be used to establish monetary values in some cases of interest to environmentalists. The general problem remains unsolved, but the importance of the ecological crisis forces us to use what little we do know as intelligently as possible in establishing rational economic and metaeconomic planning measures.

Fun for Biologists

Obviously, if we lose a lot of species of plants and animals through man-made changes in the environment, it is going to cut down on the fun for those biologists who have been earning their daily bread by writing journal articles on the species in question. Naturally, the ecologist par excellence has a special axe to grind, and rightfully so. The experts in a field should, at a minimum, be concerned with the preservation of the subject matter of the field. And since all of us are concerned with ecology, we tend to look to the biologists as the logical leaders in propaganda campaigns for ecostabilization.

But could it be that many of those concerned with the problem are too closely associated with the necessarily special viewpoint of the biologist? There can be no harm, in general, in encouraging the public to heed the advice of experts, but a specialized viewpoint on the problem can make the environmental movement particularly vulnerable to outside criticism. Cynics can attack concern with ecology as "a fad nurtured by the arrogance of scientists, nature worshippers, and radical dissidents." Who worries about epiphytic algae in the Everglades? [29] Not everyone, by any means. The environmental crisis should concern everyone; it directly affects the air we breathe and the food we eat. So it is too important an issue to be even slightly tainted by the suspicion of special pleading, by biologists, or by anyone else. Of course, most biologists and other scientists have a better idea of the big picture of man in nature than do most people. Hopefully, they will make every effort to stress reason and balance in making critical decisions.

Having said this much, we might now consider a more specialized argument of biologists on ecology that may be expressed in metaeconomic terms; we refer to the concept of the preservation of the genetic pool. Hybrid corn and many other food products have been developed by using multiple mutations of species found in nature to generate new, superior species and varieties. The trouble is that the pure species, other than the immediate parents of the hybrid, may then be neglected and eventually may disappear. This process may possibly be happening with corn, i.e., the hybrid varieties may be driving out all others.[30] If it does happen, where will new hybrids (to adapt to new diseases, new pests, and new climatic conditions) come from? There is no nonsense about an economic value here. The value is there, but it is hard to say just how much it is in a dollars-and-cents way. Even if the value (or benefit) is hard to pin down, it must be so large, over the long-run, that funds should be allocated to preserve a safe margin (99 percent reliability) of existing strains. In this, there can be no real argument with the biologists.

Most species of plants and animals cannot be treated this way because most species are useless in any marketable way. The preservation of species and habitats falls into the hazy area (in the economic sense) of esthetics, education, humanism, etc. In a human sense, the value of the world as it is arouses decided feelings of value. Some facets of the preservation problem have already been touched on in the discussion of esthetics. The preservation problem leads us into the complex subject of evolution and what our role is in evolution. We look at this next.

6

EVOLUTIONARY
MANDATES

We know how to handle planning in terms of values measured in money terms. With a bit of courage and the realization that important decisions about the ecosystems will be made whether we dare to make metaeconomic value judgments or not, we can extend the pricing mechanism to an evaluation of factors usually not connected with economics. But is there also an area of value judgment into which even the foolhardy will not venture at present? The answer, in general, is yes. More important than the affirmative answer is the fact that, although there are probably areas that cannot now be treated sensibly by any quasi-economic analysis, even if zero or infinite prices are used, future centuries may demand careful planning in precisely these very difficult problem areas. Since some of these considerations may extend beyond our concern with ourselves or our descendants, we call these areas possible "evolutionary mandates."

Do We Need the Elephants?

In his novel, *The Roots of Heaven*,[1] Romain Gary describes the struggle of a small group of men to save the African elephant from extinction. Their efforts to save the species come into conflict with the ordinary economic values of ivory hunters and even with the self-respect and desire for change and "civilization" felt by the Africans of the region. In a final defense of the elephant, one poignant reason for its preservation is given: man would be a little more alone without the elephants.

At first glance, one can say, from a cold scientific point of view, that this value placed on elephants is recreational or cultural and can be treated by the method of equivalent pricing discussed above. Perhaps this is so. Certainly, the elephant can be an economic "draw" to a circus, zoo, or safari. But we refer here to some "extra value" of the elephant. For example, in the novel the value of the elephants is taken to accrue not only to those in Africa, but also to those who never see an elephant, for example, to school children who only know that an elephant exists. In this "extra value" sense, the elephant can be taken to stand for more than a particular example of conservation practices. The elephant in this case has as a concept a very strong component of fantasy, and it has the same reality, as well as the same lack of reality, as a unicorn. So the elephant is an example of that part of the mental or possibly spiritual side of man that is not encompassed in recreation or travel or any other quasi-economic process. Neither is the elephant a free good (ultimately), such as, say, religious enthusiasm is.

In many of the dire warnings given by those concerned with ecological problems, the "elephant syndrome" figures prominently. A common ecological nightmare, for example, is an extrapolation of the traffic engineer's work to the point where the world is primarily a concrete or asphalt slab. There is room in such a world for parks and recreation, but there is no room for herds of wild elephants, roaming freely over vast savannahs.

It is hard to tell how much people value these elephant factors. This is, of course, an old issue to conservationists. If our ignorance about life processes were not so great, this type of problem could be ignored until the point where people decided they did need a wild herd of elephants (or herds of blue whales, flocks of condors, whooping cranes, etc.), and then elephants could be bred in accordance with their relative metaeconomic value. Unfortunately, the elephant is an irreversible creation, at least according to our present and projected standards of knowledge. Therefore, some point of view should be taken on general conservation practices without waiting for the value question to become tractable. Once the elephants are gone, they will be gone forever.

Up to the present time, saving vanishing species has been possible at fairly low cost. Therefore, in cases of doubt, adequate publicity and small budgets have been enough to help to preserve vanishing flora, fauna, and landscape features. In the future, economic pressures will become greater. Decisions will probably have to be made either to assign a zero price or an infinite price to such factors before progress succeeds in assigning a permanent, irreversible infinite price.

We see no magic prescription for deciding questions of this sort. What is evident is that ignoring the economic reality may make well-meaning conservation efforts into a futile endeavor. It also seems plausible, in practical terms, that some sacrifices will have to be made and that a zero price will have to be assigned to some familiar features of the landscape. At any rate, these problems deserve the rational attention of the planner in the years of the near future.

The Decaying Solar System

The world will not be with us forever. As we mentioned in the first chapter, the changing character of the sun and its expected life cycle will eventually make conditions for life impossible on the earth. In the meantime, of course, we will run into the more pressing problems of inadequate recycling of such vital elements as water, minerals, and air itself. So two very general long-range problems threaten the course of the evolution of the human race as such: first, the difficulty of maintaining our environment to permit survival here on earth, and second, the problems of transplanting the human race when eventually the earth environment becomes impossible.

In completely rational metaeconomic planning, both of these problems would be included in an overall assessment of goals and possibilities. Preparations for colonization of a distant solar system by the future inhabitants of the earth would be merely a specific project in the overall scheme of things. In practical terms, however, people tend to plan only for the relatively near term period. This tendency is rationalized consistently in the engineering economics practice of discounting future returns or benefits, so that, for example, money earned next year is worth only 92 percent of the equivalent sum earned this year. This, of course, fits in with the fact that if I wanted next year's money right now, I would have to pay 8 percent interest, say, for that period. But the problem of funding a new home for man may not become critical for many years yet. So at most reasonable rates of interest or discounting, any returns from colonization of other planetary systems as an emergency measure in the far future would be, at first glance, practically zero. If we consider returns in any ordinary finite sense, this conclusion is probably correct. But since the question involved is that of the survival of the human race, it might well be that the returns should be accounted infinite, and such a transplantation project would then make sense as a special feature of the planning.

Similar considerations may hold for some of the ecosystem problems requiring recycling of materials. This is, in principle, a purely national economic problem and is mentioned as such in Chapter 9. But the advantages and disadvantages of straightforward recycling schemes are difficult enough, and between such plans and radical solutions to world resource problems, such as transplantation, lie a continuum of vague but interesting possibilities. An intermediate-type example of this sort might be the possibility of replenishing minerals that are lost by in some way transmuting other elements on earth or by importing minerals from other planets. This possibility is certainly technologically closer to practicality than colonization schemes and might serve as a useful exploratory research project.

New Species to Replace Us

The complex nature of man makes it difficult to establish even an approximate accounting system for his wants and needs. We have seen just how difficult it is. We have tried to fit patently economic factors and external and other nonmarket factors into a metaeconomic framework in the preceding chapters. In addition, the problem of the elephants, or of a general pantheistic or spiritual feeling for nature, involves at least an important residuum of value that resists classification. We can only recognize that this last problem still exists and will have to be faced by future generations.

Future generations will also have to face the problem of survival on an earth that will inevitably deprive human beings of their present ecological niche. Again, the attitude of present generations to this problem will surely be the subject of self-examination in the near future (if there is a future—we have to remember such ecosystem disturbances as nuclear weapons). Here we can only come up with a recommendation for serious research and planning.

Both problems, while containing difficult assumptions about the attitude of man toward various phenomena in his environment and the attitude of man toward far-distant generations of man, do not include all possible value decisions for the general ecostabilization dilemma. We must consider the peculiar possibility that the future history of the world will work out in such a way that mankind cannot feasibly survive, but that life in the world may go on.

What should be man's attitude toward the survival of the biosphere without man? Or is the possibility of the survival of, say, insects and other lower forms, without the presence of man and other mammals, too bizarre a problem to worry about? It is certainly easy to say that such a world is of no interest to us, but men are complex beings, and their value preferences, as has been discussed by recent philosophers, contain definite obligation preferences (those concepts that used to be called moral values), in addition to the pain and pleasure principle. Man is mortal, but he has the distinction of behaving as if he were not. There would certainly be no probate courts if human beings had no interest in life after they no longer exist. People seem to gain a sense of immortality by controlling the actions of others after their own death, that is, by restricting inheritances through setting conditions of marriage, age, sobriety, etc. on the legatees. And people do have an abstract concern for the state of the world. Political scientists were recently tempted to postulate that older people, on the average, are less interested in current affairs and politics than younger people, since older people are less involved in the consequences of political actions. Empirical tests of this hypothesis, however, showed the opposite was true. Older people had a keener interest in the world about them than did the younger generations.

Ecological worry certainly has strong elements of this concern beyond death, even though propagandists for ecological planning have stressed, not unnaturally, the more immediate consequences of the problem. Similarly, one may well wonder whether reactions of revulsion to antiutopias, such as the *1984*'s and *Brave New World*'s, are merely the result of personal fears or actually represent an empathy for future generations. Admittedly, these examples may have been more forceful at the time of publication. 1984 is now merely a year in the next decade, while features of *Brave New World* surround us.

Descending one level down, from empathy for people to love for animals, we know that pets—dogs, cats, and even canaries—occupy important roles in the affections of mankind. Again, all of these have been recipients of well publicized last testaments. To imagine man handing over the world to a society of insects is apt to produce feelings of revulsion in many of us. To imagine a world of French poodles or a society of *101 Dalmatians* has been proved to be within the bounds of popular fantasy. Since it is difficult to know precisely what the ecological conditions would be if man had to select a successor species to himself, it is hard to know whether one could look forward to a world of dogs or not. It is perhaps reasonable that if man can transfer affection and concern to certain species, he can learn to live vicariously through other types of animals, even those now classified as detestable vermin. This is especially so if a forecast can be made of the types of species that would be most adaptable for survival in the world in the year 19,841,984.

Discussing such science fiction possibilities may appear too far removed from present-day concerns. The consideration of other species is really presented here as an example of the kind of thing that might be considered under the general heading, "Miscellaneous Ecological Problems." As long as we are at it, it might be interesting to look at some of the consequences of nonhuman survival for how we plan our lives during the next few millennia. First of all, it might be a good idea to establish small research programs, if not now, at some time in the future, to identify what species of plants or animals might be capable of long-range survival. Some consideration has already been given to this problem in connection with forecasting the environment of a post-nuclear-attack world. Then, after the candidate species have been identified, there might be some interest on the part of human beings in imparting some of the knowledge or other characteristics of human life into these other species, insofar as possible. For this purpose, training zoos could be set up, and new generations of appropriate insects or algae or horseshoe crabs might be bred to produce desirable characteristics. Cytoplasmic heredity and training would find an exceedingly interesting application here, at least from the point of view of scientific investigation. Presumably, a special educational program could be set up, aimed at establishing empathy between man and his successor species—although perhaps not so much empathy as to cause separation anxiety when the horseshoe crabs must bid a last farewell to mankind!

Far-out possibilities undoubtedly are abundant for a very general treatment of ecosystem stabilization. Contenting ourselves with just the few instances mentioned, we will now turn to a consideration of planning for man on a rational metaeconomic basis in the relatively near future.

7

POLICY ALTERNATIVES:
THE MEANS TO AN END

Economics is a positive instrument for improving welfare. If we can describe an economic system, the principles of the science of economics can help us to solve the ecological problem. Our major interest is in the application of economics in the context of the broad sweep of the principles and problems of policy alternatives. To be sure, there is a large number of policy alternatives, but which ones are best for the resolution of the ecological problem will depend on the nature of the means to the end we define. An end may be good or bad by ethical standards, but whatever our values, the present economic success of our system should hopefully provide means to help us achieve the end.

In summary, there appear to be four broad policy alternatives which may be considered appropriate means for the ecostabilization of the present system. First, and probably not so obvious, is the "do nothing" policy. Certainly, it is an option that man can choose. That is, by pursuing his own interests, whatever they may be, he can simply ignore the externalities associated with his activities and let nature take its own course. Of course, man may be terribly uncomfortable with nature's solution; we will weigh the possible disturbing results of this policy in Chapter 11. Second, an ecological disturbance tax equivalent (or very near) to the value of the externalities generated by producing and consuming units can be placed on these units. That is, by bringing the disturbed ecosystem to the marketplace, producing and consuming units pay for the privilege of disturbing the ecosystem. Third, ecological standards can be generated based on human and nonhuman criteria, and laws can be passed to dictate the movement of the system towards ecostabilization. Fourth, the forces of opinion can be pressured by various communication media to direct the ecostabilization process at a personal level. As in China, where every man's duty during recent years was to kill a fly, the forces of opinion can direct every man toward a decontamination exercise. Such exercises can be successful. There are (at least, officially) no flies in China. A well-known Indian demographer, while touring China, was deeply impressed with the concentration of the Chinese people in pursuing their "no fly" policy. Every man, woman, and child was issued a red fly swatter which, he maintains, they used with extreme vigor. While lunching one afternoon, a fly presented itself to partake of the demographer's meal. The demographer dutifully reported this rare appearance to the proprietor. The proprietor, using his red fly swatter, brought to a quick end the gourmand activity of the fly. Extremely pleased, a smile of triumph on his face, the proprietor bowed and informed the demographer he had witnessed the killing of the last fly in China. It should also be noted by the reader that there are no sparrows in China, but that is another true story.

Pricing and the "Invisible Hand"

By now, even the mere watchwords of the economist—supply and demand—should be enough to suggest to the reader the vast array of underlying cybernetic forces that determine the price of a good or service. In a market system, the interaction of buyers and sellers will determine a price and the quantity of the good or service exchanged at that price. Certainly, one policy alternative is just to allow market forces to continue their operations as usual, so that the general equilibrium of the overall economy is maintained. But externalities are not fully accounted for in the ordinary pricing scheme we see in operation in supermarkets and stock exchanges. It is only when products associated with the disturbed ecosystem are included in market calculations that the price mechanism allocates resources for regeneration. For example, household and production units purchase water in large quantities for their own use: cleaning, washing, cooling, sanitation, irrigation, etc. Costs are reflected by the facilities that provide this service—the complex of dams, waterways, recycling plants, and so on. Whenever this water is contaminated and channeled for deposit in rivers, lakes, oceans, etc., there is another cost associated with the contamination of the depository. When a bull elephant is shot as a sporting thing (or more likely, as some sort of effort toward the psychological gratification of man's earlier nonhuman attributes), it affects the pleasure of other individuals in viewing or appreciating the existence of pachyderms in general. When one individual discards beer cans in an unprotected open field, others' enjoyment of that field is less. These are all forms of negative externalities (or external diseconomies) associated with the given unit's use of some resource that affects other units in a negative way.

The difficulty is obvious. Only when producers are willing to supply and consumers are willing to pay for open fields, pachyderms, clean rivers, and lakes, will Adam Smith's invisible hand come into play, still assuming such products can successfully be brought to the marketplace. In technical terms: When supply and demand functions do not intersect there is no price. Demand may be such that buyers are not willing to pay prices as high as the sellers insist on for the smallest amounts they are willing to sell. For example, entrepreneurs may be willing to supply pachyderms (clean rivers, etc.) but their costs for doing so is high. Buyers are not willing to pay the corresponding price. Should costs appreciably diminish, or demand increase through an increase in desire or incomes of individuals, then a price would result. Another situation exists when sellers are willing to let buyers have some amount without any charge, and the demand of buyers is such that they take some amount less than being offered. Hence, there is no price, and the product is a free good. If demand increases, say, for viewing and photographing elephants, or the supply of elephants decreases because of the entrepreneurial activities of ivory hunters, a price will be established. This is the world of Adam Smith where the desires of man, as a consumer who votes with his dollar, interact with the profit-seeking of the producer who attempts to meet his needs.

The economic record indicates so far that gaps remain between the supply and demand schedules for some products, i.e., there is no interaction and no price. Even when there is a price, it may not always adequately reflect the externalities associated with production and consumption programs; there is not an accurate social weighing of their alternative costs. Man has been willing to allow the disappearance of several species, continued pollution of his waterways, and scarrings of his open fields. Of course, man is human and entitled to make a few mistakes, but it is probably too cynical to think that anyone who is not a lunatic in this post-Freudian world would not want to place some democratic restrictions on man's behavior. So in our era it is an acceptable fact that the representative government of an effectively participatory democracy can and should be stimulated to provide some of those actions that can lead to a stabilized ecosystem.

The Government as
an Economic Determinant

Our system mixes government and business. Noisy controversies about "government versus business" can mislead by focussing on elements of rivalry between the two sectors. There never has been a complete agreement on the best dividing line between the two. At a minimum, we have used the government to do such things as to set certain general rules of law in order to secure order and internal peace, and for the judicial settlements of disputes; to define weights and measures, and, in general, to fix a framework within which we can carry on our affairs. Today, however, Americans do far more through government and political decision making, and, incidentally, pay far more for the privilege. For example, governments carry on hundreds of different activities: health inspection and regulation, the operation of bus lines, a railroad in Alaska that is making a tidy profit, the management of large electrical utilities, and many social activities once left to charity. Thus, over time the community has shifted to government functions that are not traditionally governmental. Trying to analyze the efficiency of the resulting system would be tedious, partly because the system is so thoroughly mixed. The auto industry seems efficient and progressive. It might be more so, and cars would certainly be cheaper, were it not for taxes. Yet without vast government spending on highways and streets, what would our auto industry be? More likely it would be a pygmy and certainly not the giant it has become. So it seems natural, if not inevitable, to suggest the government as an economic determinant to resolve the ecological problem.

Experience suggests that we are more willing to delegate jobs to the government than to provide funds to do them well. Perhaps there is a form of diminishing return so that,

131

beyond some point, the more jobs we try to do through government, the worse we do each. Yet new techniques may permit us to escape old limits, e.g., the calling in of business firms to operate government installations, notably those of the Atomic Energy Commission, and the consequent freeing of public officials from what are heavy duties of management. A decision to use government to help stabilize the ecological system still leaves open the question of selecting which level of government. Certainly there has been a tendency to rely on the national government since the depression of the Thirties and the war of the Forties. Many ecological problems are not predominantly national and policy could be usefully adapted to varying local needs. At the state level costs could be identified more closely with gains, and inefficient policy could be less easily foisted on the public because costs are not concealed or apparently borne by outsiders. This theoretical advantage, of course, presupposes a public visibility that is often regrettably lacking in actual state governments.

A further point is that competition among states could tend to protect the freedom to check the abuse of power in any one state and to limit the evil that a mistaken minority could accomplish. The danger of mistakes with widespread consequences is naturally less if decisions are made independently by several centers of power. Unfortunately, these "competitive" advantages of decentralization have often been more theoretical than real. California and Arizona are no inspired examples for the benign uses of competition in water policy.

The national government, on the other hand, occupies unique vantage points for supporting research and development programs, and the federal income tax has proved a uniquely effective way of raising funds, while state and local governments have moved from one tax crisis to another. So the national government has developed as a dispenser of funds to other levels. It could and does aid communities and states by setting up matching funds to support their programs. The size of this expenditure is directly related to the alternative costs involved; that is, what must the government give up to support positive ecological activities? One opportunity for scientific economics is to search for possible areas of conflict and cooperation, and especially to look at the long-run implications of these policies.

If desired ecological objectives have values that exceed their alternative costs, programs can be initiated at the local, state, and national level. If the priority value of resolving the problem is high, then government may act as an economic initiator in resolving the ecological problem. The degree of its participation, of course, would depend on the gravity of the ecological problem, and in pathological cases it could conceivably require a very high level of positive economic activity, perhaps even comparable to the combined expenditures for space exploration, the Vietnam War, the Korean War, the Second World War, the First World War, the Civil War, the Revolutionary War, the War of 1812, and the development costs of the Tennessee Valley Authority.

Government can also use "cost-free" economic means. An interesting possibility for the use of government power in solving the ecological problem is the imposition of compensatory taxes (or "effluent charges"). This type of taxation has been used very little, except in such user charge contexts as the California gasoline taxes, which are used to pay for freeway construction, but it is probably something that should be tried to help solve the present impasses in the environmental disruption problem.

If producers or consumers are taxed to compensate society for the environmental damage they do, many problem areas that cannot be treated by regulation will become amenable to control. While regulation is often all or nothing, taxes can be flexible. Also, regulation can be either effective or ineffective, depending on public opinion and efficiency of enforcement, but taxes either work to help control pollution and other diseconomies, or if they fail to do that, they at least produce income (money) to pay society as a whole back for the damage. In addition, the taxes collected can be earmarked for direct government aid to the environment, if desired. So we consider taxation a powerful but relatively unused weapon in the campaign against environmental disruption. We return to its use in Chapter 9.

The government, therefore, has many economic policy tools available through its powers to tax and spend. Of course, there is also the standard role of government, the accepted political powers of regulation and control, of which more below and also later, in Chapter 9.

Legislation and Control

Rule setting has been a primary activity of our government. Bodies of regulatory law, enforced through our courts, are essential for the operation of our mixed economy. Of course, bad laws or bad enforcement will add to the difficulties of the working of any society. As a possible alternative for resolving the ecological problem, legislation and the proper enforcement of that legislation is a plausible policy alternative.

Congress should, ideally, take the initiative in enacting legislation associated with a disturbed ecosystem. But what kind of legislation? Certainly it could consider bills for each and every facet of disturbance in the ecosystem. Alternatively, it could declare that it is a continuing policy and responsibility of the federal government to use all practicable means to foster and promote the general stability of the ecosystem. It need not specify what has to be done if a particular ecosystem disturbance develops, but it should provide for agencies to help it; the President and the public decide on what course of action to follow. In the process it could at least create a joint Senate/House committee to study ecological reports and establish a small professional staff to help analyze ecological conditions and proposals. A somewhat more ambitious scheme is outlined in Chapter 12.

Any ecostabilization legislation and its associated practices may run into obstacles. For example: (1) our economy/environment is so extremely complex that what seems wonderful for one part may be all wrong at that time for another part; (2) we cannot be very accurate in regard to such legislation in predicting what will happen; (3) various interest groups will certainly oppose forms of ecostabilization legislation that hurt them specifically; and (4) at present levels of the public understanding of the ins and outs of ecosystem disturbances many persons whose intentions are excellent will oppose desirable programs and even work for policies that are dead wrong. Perhaps what is needed is broad educative action at strategic points.

In addition, there is the enforcement problem of such ecostabilization legislation. Regulatory hearings may back up for months. Defendants will naturally use any legal posture that can be reasonably employed and financed. Therefore, often only long-term results should be expected through the law. On the other hand, acceptance of and compliance with ecostabilization legislation by both producers and consumers could result in a positive alternative policy. But such a happy result probably requires the moral pressure of well-organized public opinion measures.

The Forces of Opinion

Another policy alternative associated with an ecostabilization program is the use of the forces of opinion. Such a policy would act to make society as a whole believe that continued consumption and production trends would result in a negative human environment.

135

The generation of such opinions has been associated, historically speaking, with the efforts of a few individuals. For example, Karl Marx pursued an idea that resulted in the development of a "new" society; Calvin developed an idea that became an ethic; even Ayn Rand has a following. More than the development of an idea, there has to be a public that is generally willing to accept and translate these ideas into concrete programs. Groups and organizations develop, publications evolve, and the idea spreads. How far it spreads depends to a large degree on the dedication and activity of the participants; like a redwood, ideas take time to grow. Opposing pressures, the breakdown of traditions, and old ideas and habits make opinion change difficult. As an alternative policy, public opinion manipulation warrants special attention, because all the alternatives developed above are in part directed by the forces of opinion.

Positive opinion toward the disturbed ecosystem can be generated in many ways. Through the many communication media nowadays, interest in the problem can be easily established in the public mind. Again, experience dictates that unless the externality is a personal one, opinion remains relatively dormant. Even if it is personal, opinion may be lukewarm. For example, the American Cancer Society has publicized what they believe to be the deleterious effects of smoking. We are sure that most smokers are impressed with their findings but continue to ignore their warnings, due in part to utility or satisfaction associated with smoking. A mass media program (like that addressed to smoking) could be undertaken to show the deleterious effects of chemical emissions from continued automobile use. Although such facts are agreed on and accepted, most people simply will not give up driving automobiles. If in protest to present emissions technology, people did at least try to cut down on unnecessary driving, Detroit, we speculate, would undertake an R & D program on a very large scale, to say the least. The generation of action so far has been in the form of putting responsibility on gasoline and auto producers for resolving the problem, and given the structure of these industries, we would expect just the kinds of pollution "marketing games" that have, in fact, recently occurred.

Even personal experiences of eye irritation, or of witnessing the feeble movements of an oil-soaked bird, may not measurably reduce apathy. The losses of standing timber to diseases created by smog, the disappearance of the blue whale, and the contamination of our beaches and rivers are all of public concern, or interest, but the force of opinion necessary to resolve this imbalance in alternatives is often lacking. Again, man is an adaptable animal, and he is adapting to his new environment. To be sure, he is opinionated, but, unfortunately, he often lacks the force to stand by his convictions.

The Policy with No Name

The ecological problem is international in scope. Certainly various and sundry policies can be pursued for resolution of the ecological problem on a purely national basis. Positive action on the part of the government, producers, and consumers could create an example in one nation of the potential benefits of intelligent policies. Nature, however, acts and reacts within a total world system. How is this total problem going to be resolved? As an alternative policy to attack this problem, we can suggest only a diaphanous creature that perhaps deserves no name.

What is ideally required here is the development of a strong international organization of qualified individuals backed by the support and legislation of a supranational controlling agency. Given the nature of disturbed ecosystems, as opposed to military defense systems, it is at least plausible that international cooperation along these lines could open paths of noncompetitiveness, or better, direct competitiveness towards a common goal. Science perhaps points the way, as the Pugwash Conferences attest. Or perhaps economics itself can help, as with the growth of the common market concept from Europe to Latin America and Asia. Any evidence of the growth of cooperation on a nonphony basis is encouraging. A recent agreement between President Nixon and Japan's Prime Minister Eisaku Sato for an intensified program of cooperation in solving mutual air pollution and broader environmental problems is an example of what we mean. Naturally, the difference in the needs of various nations must be recognized. The government of a country rapidly progressing from an underdeveloped status may well think that almost any environmental damage is justified to avoid losing developmental momentum. The outlook certainly looks dimmer and dimmer the more we investigate possible complications.[1]

It has been said that peaceful pursuits need to be billed as the moral equivalent of war. Certainly, a war with the inhabitants of the Epsilon Arcturi solar system would unite China and America in a common effort. It is too bad that pollution and other environmental problems cannot raise the collective adrenalin of the nations of the world to similar cooperation.

8

SOME TOOLS OF
ECONOMIC ANALYSIS

After the money or supermoney values of economic and non-economic factors in the ecological crisis have been determined to our satisfaction as adequate for planning purposes, and after we know what policy possibilities we can follow in order to adapt the market system to produce ecostabilization, we must create practical plans. To produce a practical dollars-and-cents plan, a mathematical framework must be set up to assist us in complicated decisions. In drafting this mathematical structure, various tools of economic analysis are important. We talk here briefly about the role of market analysis in the ecostabilization problem and also about methods for setting up the optimum or "best-decision" equations for policy decisions. We also sketch out the role of statistics or uncertainty in making reliable decisions and make some mention of value transference over future time.

To Market, to Market

Market analysis consists of assessing and evaluating the structure and behavior of the market. Associated with this structure and behavior are various degrees of industrial concentration. For environmental purposes we must worry, at a minimum, about the nature and significance of ecostabilization leadership by enterprises, the influence of market structure on technological progress and ecostabilization, and the public policy within that market.

A firm working within an ologopolistic framework has a price policy determined by the existence of rivals whose action it must take into account. This relates to the standard gaming analysis discussed earlier, in which the firm weighs the effect of present prices on future prices, the possibility that its price in one market affects its price in another, or the possibility of competing in ways other than price, and generally through many other options. In order to do this, some degree of market control on the part of the seller or the buyer (oligopsony) [1] is required. Of course, these types of policies may or may not reflect responses to a disturbed ecosystem.

In those sectors with low levels of concentration of firms, e.g., lumber and agriculture, firms pursue marginal profit goals where their output decisions depend more closely on the prices being determined in the marketplace, that is, they are price-takers rather than price-makers. Increases or decreases in output are due to price changes beyond their control, and may or may not disturb the ecosystem. Therefore, it is important to properly assess markets not only in terms of price-output behavior, but in their contribution to disturbing the ecological environment. This leads us to the important question: Given the general reciprocal nature of supply and demand, can firms and their policies resolve their contribution to the ecological problem in light of their price and/or production policies?

The nature of leadership, that is, the dominance of one firm within a given industry, is an important consideration in market analysis. For our purposes, we must try to estimate the willingness of the dominant or "barometric" firm to accept the responsibility for its contribution to disturbed ecosystems, and we must determine whether or not there is a correlated positive response on the part of other firms within the industry. Since its policies are closely tied to the market, certain rewards and punishments arise naturally within the industry as a result of the actions of a given firm. In other words, if the firm believes it can increase its profits and share of the market by pursuing an environmental program, it may well do so. For example, Royal Crown Cola has recently initiated a program of bottling its beverages in old-fashioned glass bottles, advertising money deposit requirements as an inducement for the individual to return the empty bottle for recycling. Certainly, it is an interesting way of saying, "Drink our cola and save the environment"; but other firms are beginning to pursue the same program which could well compound into an important environmental program.

It is important also to consider in this ecological market analysis the willingness and ability of individual firms in a given industry to undertake research and development programs for decreasing their contributions to pollution. Once again, through reward and punishment significant applicable steps toward ecostabilization may be made. Inventive activity is stimulated because a reward is made for decontamination inventions. Obviously, a patent can encourage inventiveness by promising temporary ownership rights to the invention. Ideally, each new method should be made immediately available to any interested firm. But in practice some firms acquire effective monopolies on ideas. So it may be that R & D development does not favor atomistic competition, but at present it is impossible to form an opinion of what firm size is favored in various industries. So environmental economists must be prepared to face various classes of reactions from various types of industrial markets.

It is also important to assess the effect of governmental policies on the market, since pursued ecostabilization policies will affect the nature and workings of that market. If the market is not economically evaluated, haphazard policy decisions may very well result in further deterioration of the ecological system along with disastrous economic effects.

We can summarize the background for market analysis by viewing the marketplace as made up of monopolistic firms at one pole, competitive firms at another, and a continuum of oligopolistic or "monopolistically competitive" firms stretching between them. For environmental planning purposes, the important thing to note is that at the monopolistic pole the firms have considerable power to change prices or products. Therefore, we can propose enforcing regulations or taxes for ecological purposes on these firms with the knowledge that they have some freedom of action. They can vary their activities by changing products to reduce pollution or by contributing money to repay society for the costs of the pollution they cause. At the other, perfectly competitive pole, little or no flexibility exists, and we have to face the fact that changes to the economy will inevitably cause dislocations and temporary losses in productivity. For the intermediate continuum of firms, we expect intermediate flexibility. All these possibilities will be touched on later in Chapter 9.

We now want to set up the following requirements for a satisfactory ecostabilization analysis: (1) It should be historically related to the ecosystem in the sense that technological, organizational, and public policy influences have produced the present system. (2) The analysis should be able to explain why the market is as it is in disturbing the ecosystem. (3) The market analysis should be as dynamic in nature as possible because market structure is potentially variable in its effect on the ecosystem, that is, those variables selected and acted upon may set in motion a chain of reactions throughout the industry. (4) The analysis should go beyond explaining the cost-price-output relations for a given product and must be prepared to appraise conditions of choice among other variables and the economic result of that choice. The above list is

142

formidable, and we must be modest in our demands for accuracy but obdurate in demanding that the best possible analysis be made. Otherwise, much economic damage may be done without necessarily helping the ecosystem at all.

Weighing the Pig

Whether weighing pigs or planning smog control, the best answers must be found. Nowadays, we use the rather elaborate name, "optimization procedure," for the process of finding the best plan. Finding the best plan for a community or a world is, in principle, no different from finding the best plan for a business firm. A business firm wants to make as much profit as possible. To the best of his ability, the businessman tries to marshal his resources—his capital, labor, and land—in such a way as to make the most money possible. The community and the world will essentially want to carry out the same process, if we enlarge the concept of the word "profit" to include the satisfaction of various human values not usually entered on the balance sheet.

Every businessman wants to know how to plan his activity with a profit goal in mind; that is, he wants to carry out an optimization procedure. Usually, both businessmen and governments learn what their optimum plan is only by trial and error. Naturally, it would be nice to know all of this beforehand. Often the consequences of an individual string of events are known, but the overall pattern of possibilities is too complex to figure out. At this point, the modern science of operations research, which amounts to a more or less rigorous use of mathematical equations in logically solving complex problems, is of great help. Operations research is, therefore, really only an extension of the old apples and oranges problem of sixth-grade arithmetic to the problems of business, government and the world at large. In later, more universal and enlightened incarnations, operations research has been elevated to the status of a way of life and is usually referred to as systems analysis. Properly used, in any form it is a timely blessing to the pressing needs of humanity in the electric age.

To use the "cool" (in the McLuhanesque sense) modern methods we must express all factors of a problem in terms of a mathematical logic. At present, we are interested in the means of implementing this process for ecosystem stabilization. Since no such logical representation can be perfect, since some factors are necessarily left out, and since the logical relationships are often unknown beforehand, no such setup of mathematical logic is perfect. Applied mathematicians often term the process we are talking about as the optimization of a model, to suggest the relationship of a toy to the real thing. For planning purposes, however, a toy may be a lot better than the wild guesswork that often passes for expert opinion, so the mathematician need not feel bashful about his trade.

Once the model is developed, the problem is to find out exactly how many apples and oranges there are. Unfortunately, applied mathematics itself is merely a finite creation, and it is not very easy, even with the problems of a small business, to find answers through clever tricks of ninth-grade algebra or its equivalent. In principle, there is always a difficult but general way of solving problems, and in practice, there are many methods of finding good approximations. In addition, there are many useful cases in which simplifying "mathematical tricks" are available and are practical to use.

The difficult way of solving the optimization model problem is called "simulation." The word simulation has become generally familiar from television descriptions of the space flights of the recent years. Just as a simulated space flight goes through all the motions of a particular flight by having instruments read as if they were presenting data on burn, roll and pitch, various malfunctions, and so on, the use of simulation for economic purposes (for road building, for example) specifies price changes and changes in the amount and direction of truck shipments, as if they were caused by actual improvements in a road network. In the most general case, many different possible logical outcomes are possible, just as various malfunctions are possible during a space flight. Therefore, no one simulation is, in general, correct, and the process of determining a solution involves repeating the calculation over and over until some answer that gives a best result or a most probable result is obtained. In the sixth-grade arithmetic example of apples and oranges, simulation amounts to finding out the maximum number of apples and oranges one can buy by trying out many possible combinations of purchases at the market stall.

In principle, simulation is the best way of doing things. In practice, even with high speed computers, the number of different possibilities is so large that the whole age of the universe would not suffice for the calculations, so simulation often proceeds on the basis of very simplified models. In our analogy above, such a simplification might be to call both apples and oranges fruit.

So simulation is the sole method of solving the general optimization problem, but there are several useful methods of solving some special mathematical problems that occur in planning in practice. Key among these is a procedure called "linear programming." Linear programming is a useful shortcut to solving problems that involve a complex set of choices or possibilities among elements that are themselves rather simple. A typical use of this technique in the "microeconomics" of business firms is to find out how to get maximum profits in a situation of fixed prices and fixed unit costs. The elements are conceptually simple, the complexity coming from the choice of amounts of labor and goods to put into each activity that produces a profitable product.[2]

Farm management provides a homely example. A farmer is able to grow wheat, corn, and oats to feed cattle and chickens. How much should he grow or raise of each? The problem for the farmer would be trivial if the only constraint on his profit were the amount of capital he is able to put into the operation. In that case, he will naturally choose to put all of his capital into the business activity that has the largest ratio between return and investment. We know, however, that many farmers will grow a variety of crops. In some cases this is because the farmer does not behave as an economic man, as when his wife or daughter raises a few chickens in the yard as pets or novelties. Or he may grow different crops just as an insurance against selective disaster, such as the corn borer. But such diversity in farming often has a basic economic purpose, for in the general situation, not only is the farmer's capital limited, but he often has land of different qualities in relation to the possible crops grown. Another complication might be that different amounts of labor are needed at prices unrelated to the particular crop grown. Furthermore, some crops may be a part of others, as in the input-output table of the economist. An example is the feeding of corn to cattle or chickens. We can see that the problem of choice for the farmer can become very complex. This problem has been successfully treated by methods of linear programming.[3] The actual method employed saves time over simulation because in simulation all possibilities for employment of the scarce factors of capital, labor, and land are always in the running to produce the best total profit picture, but in practice only a small set of such combinations or "decisions" are not ridiculously wasteful. Furthermore, it happens that starting from any one small set of factor-use decisions, there is a way to try better and better decisions without backsliding, so that the best decision pattern (optimization) is reached efficiently. Linear programming makes use of these tricks to solve problems, without running the computer from now to the next century.

Linear programming is a useful tool for economic analysis and planning, even when things are not very linear, e.g., when returns to scale are noticeably decreasing or increasing. In such cases, one can sometimes use a mixed system of variables which may take on whole number values, or one may use others that have continuous values. That is, a "jump" from a farmer's not raising pigs to raising pigs is a nonlinear process. But if we call the nonraising "zero" and the raising "one" and allow only those two numbers, we can make the problem "almost" linear, as compared to a "real" linear problem where 1/2 or 1/3 would be a possible answer. A similar problem might come up in environmental problems, when the state is faced with making motorists spend $100 on a catalytic afterburner or letting them save the $100 at some cost to the purity of the atmosphere.

So linearity can often be "saved" in some sense, that is, by dealing with whole-number quantities. Unfortunately, there is no denying that nonlinearity is the way most things are in this complex world. Techniques have been developed for so-called nonlinear programming, and practical calculations exist for cases in which nonlinear functions have certain mathematical characteristics, such as being in the form of squares of numbers (quadratic) which can be converted into a quasi-linear problem.

Another method for computing complex optimizations is that of "dynamic programming." Dynamic programming involves following a prescription for enumerating possibilities one by one, perhaps many at some beginning (or ending) point, but then zeroing in slowly but surely on the optimum solution without examining each possibility.

The method makes use of the fact that the best answer to a problem is often the sum of the best answers to a part of the problem. For example, choosing the best route to drive to work is complicated in general, but if we knew the best route, each segment of that route would also be best (shortest in time). Obviously, such a problem is still complicated, but knowing that one is always dealing with "partial optima" often helps a lot.

As the name suggests, dynamic programming is exceedingly useful for looking at processes which change in time. Unfortunately, severe restrictions have to be placed on the use of dynamic programming in practice. Solving for too many different types of decisions becomes as complicated and longwinded as carrying out a simulation in the first place. The method, nevertheless, is of practical use in many cases.[4]

These techniques by no means exhaust the store of mathematical techniques used in economics, but they are widely-used methods and procedures, and some of them are illustrated in the environmental examples in Appendix C.

Which One Stayed Home?

Another factor to be considered in economic analysis is uncertainty. No one has a crystal ball that can tell him about the future. Individuals and businesses adjust differently to the changing world, and as they adjust they produce new change. The fact of uncertainty makes it difficult for business to correctly adjust to environmental conditions as they develop, and in making their adjustments, many months or years may be necessary to work out the long-run equilibrium. Meanwhile, conditions continue to change, and new decisions are made, and so on. Therefore, the survival principle is based, to a large degree, on making good decisions. Moreover, ecological economic activity is undertaken for the future and, therefore, involves some commitment for the years ahead. It must be guided, then, by forecasts of that future, and a good rule of thumb is that the more distant the future for which a decision is made, the greater the uncertainty. If we know what creates uncertainty, then we have added to our knowledge about what creates the opportunities or alternatives for eco-logical-economic system regeneration.

Rational decisions concerning ecostabilization programs under conditions of uncertainty must rest on the calculated probabilities of future prices and costs or, finally, on the probabilities of future benefits. We need to point out here that there are methodologies for this kind of decision making, and for the most part they incorporate the concept of cardinal utility. For example, some progress has been made in the last few years in the logic of one-person decision making under uncertainty.[5] And Von Neumann and Morgenstern have considered a decision maker who must choose among various prospects, when some prospects are available with certainty and the others constitute all the probability mixtures of the less sure prospects.[6] Future developments in this type of "game theory" may produce extremely useful results for econometric analysis.

Actual stochastic (probabilistic) planning decisions usually use much less sophisticated methods. The planner is often grateful just to have a "probability distribution" or a picture of how many times each kind of result occurs in so many tries. Poker players are familiar with the logic of "distributions," the betting value of hands depending on how seldom they are likely to occur. A full house, therefore, has a lower probability in the distribution than two pair, for example.

A statistically oriented planner would want to know as many of the characteristics of the probability distribution as possible. But in the real world, planners (statistically oriented or not) have only educated guesses and hunches about these characteristics. These types of estimates are generally suffi-cient only to establish a notion about the range of possible outcomes and, on rare occasions, about the most probable outcome.

Planners are especially interested in the range or the ex-treme values of the distribution. These extreme values would mean great losses or, alternatively, great gains, and it is safe to assume that the majority of planners want to avoid great losses. On the other hand, a minority of planners (the gamblers) might strive for great gains in spite of, and sometimes because of the possibility of great losses. Both groups, therefore, will focus on the range of the distribution, even with differing goals in mind.

Some of the more important uncertainties confronting the planner are those associated with changes of technology, consumer demand, social institutions, and nature. To repeat, the uncertainties of these changes increase with time. Conversely, the uncertainty of the expectations of a specified future date decreases as this date is approached. With this relationship between uncertainty and time, it is possible to consider the effects of uncertainty as an economic force influencing ecostabilization decisions. In other words, the planner can see how a change in uncertainty expectations affects his decisions—whether the results are positive or negative.

The role of uncertainty in human affairs also enters into economics through the use of inductive reasoning to make up new scientific laws. By "new scientific laws" we mean causal relations between concepts that we did not know were related. Of course, all it takes to make up a new scientific law is to guess at a relation between two concepts (make a hypothesis) and then get data to try to match up the facts and the theory (verify the hypothesis). If the hypothesis is verified, we have a new law, otherwise, too bad!

But nature (and man) rarely makes things this simple for us. Usually the hypothesis is true sometimes and false other times. That is, for some data we can often use a straight line to give a linear relationship (meaning the concepts are proportional) between two concepts, but other data would require a different straight line. What is usually done in this case is to try to pick a straight line that is almost right for a whole group of data. This is where statistics, or the science of uncertainty, comes in. For example, if we tried to establish the relation between drunkenness (as measured by walking a straight line) and the consumption of alcohol, we might guess that there is a general tendency for drunkenness to be proportional to alcohol. But the individual tests we observed would all show somewhat different relationships. In this case, we would try to draw a line that would give the best average relation. Which line is this? One answer is to draw the line so that the sum of the differences (squared) between actual data points and the line is minimzed. In intuitive terms, this means making the line as close to as many points as possible. This process is called "linear regression analysis" and the actual line drawing is carried out by solving a mathematical equation.[7]

In practice, the relationship between theory and fact can be improved by including more concepts (or variables). In our example, using both body weight (fat people can drink more than thin ones) and time between drinking and testing would obviously give better results. To resolve the ambiguity between the observed data points and the theoretical "law" we want to establish an average "line" (really a line in several dimensions, or a "hyperplane"). The actual line we draw can be chosen in the same way, so as to minimize the average distance of the data points from the line. Several equations, then, will be needed to actually find the best line, and the process can then be called by the impressive name of "multivariate linear regression analysis."

It is often especially hard to separate out the effects of pollution on, for example, recreational values. So these statistical methods are very important in environmental problems. In Appendix B, we discuss a water quality problem that involved using these methods.

Roast Beef in the Future

Man lives in the present, but worries about the future. In planning for the ecological crisis, we must not only concern ourselves about the economic and metaeconomic values to be given to goods and services at the present time, but we must also decide the relative worth of things in future years to come. We want to produce the best result, as far as the state of art is concerned; however, we must recognize that some compromise must be made between the results that we wish for the future years and the results that affect our present lives. In theory, one could make any kind of choice, assigning all importance, say, to five years from now, while ignoring the present entirely. Usually, such an arbitrary way of doing things is not a popular approach. Some kind of thinking allied to this point of view may go on, it is true, in the setting up of "five year plans." In that case, it is also true that the five years represent, not a purposeful putting off of results desired, but a necessary length of time in which to effect the changes desired. In addition, there is always a tacit assumption that the results in the sixth year and following years depend on the state of the generalized economy in the fifth year.

At any rate, in a mathematical formulation, we must decide how much weight to give to the profits for society, from an ecological standpoint, in each year of the future. Economists customarily treat this problem by supposing that people always prefer present profits to future profits, and future profits are less desirable as time goes on. This method of analysis is called "discounting" and involves the division of future profits by a number depending on a discount or interest rate. Indeed, the discounting procedure can immediately seem plausible when considering the interest paid on money. If I paid 8 percent interest on a bank loan to be repaid at the end of one year, it is reasonable to infer that the money right now is approximately 8 percent more valuable to me than the money next year. So insofar as the interest rate actually represents the possible earning value of the money, it is a justification and a model for the discount rate we need in planning. In practice, we caution that the appropriate rate to use may be a matter of controversy. Actual rates will depend on the imperfections of the capital market and, of course, on the amount of risk or uncertainty in proceeds from the investment.

Once we have decided on the discount rate, we can compare the returns (usually called benefits) from our policies for each year and the corresponding costs. One way of combining total benefits and costs is to add up the total discounted benefits and costs for each year, producing a net present value of the total "stream" (or year-by-year sums) of benefits and costs. By comparing the ratio of these present net values, one can see whether the returns are greater than the costs and, if so, by how much. Other methods differing in detail may be used, such as computing a discount rate from the arbitrary equating of the benefit and cost streams. The best choice of method is a question of controversy, or perhaps we should say, taste. Of course, we have been talking in usual terms of engineering economics; for treatment of the ecological crisis, it must be understood that the benefits and costs include both the economic, and what we have called metaeconomic values. Discounting is used in the difficult problem of the evaluation of health benefits mentioned in Appendix B.

We have assumed above that present profit is preferable to future profit, or that discount rates are always positive. This assumption follows general practice in the evaluation of planned projects, but one can see that in ordinary life, this may not be so. A rather familiar example is that of Christmas Clubs, in which people save money for which, traditionally, they have often received no interest. Here the bank is acting

153

as a depository of funds that one would rather have at Christmas than in March or April, and the zero interest rate partially reflects this predilection. Similarly, a student might need several thousand dollars at some future date in order to go to college. It is possible that money will actually be worth more to him at that date than it is now. In that case, the interest gained on money saved now in a conventional bank savings account to go to college is actually an overpayment to him personally, or you might say that he has acquired a consumer surplus of a kind. We might expect this type of case to occur in ecosystem stabilization problems, because as society grows more affluent in the future (if it does), the utility of consumption of recreational values might become higher, and, so the discount rate associated with recreation would fall. Similar problems can come up as far as the value of capital as a production factor is concerned. Economists have thought about this [8] and the difficulties are not insoluble. Even if the configuration of the world at present and future times turned out to be such that negative interest rates would be indicated, planners would not be too dismayed. At any rate, one must keep an open mind as to the relativity of present and future values in such intricate problems as the analysis of the ecological crisis.

9

ECOSYSTEM
REGENERATION

The ecosystem can best be regenerated by utilizing all public
and private means at our disposal. Typically, public planning
will play an important role. We have seen how the science of
economics can be utilized to help decide how to plan most
effectively for our survival as a world. Now we might look at
some general examples of ecosystem planning.

First, let's look at the general rationale and the key elements used in planning; then, we will go on to how a government commission might go about the general problem of
choice and policy in planning for ecosystem regeneration.
Some specific examples of the use of mathematical planning
techniques are discussed in Appendix C.

Making the Best of It

We have emphasized the central role of the market economy
because we believe the rules of the market reflect the basic
truth that you can't get something for nothing. Clean air for
our smog-troubled cities, for example, would cost us some-

thing, and the significance of this cost could best be imagined, not by thinking of abstract dollars, but by thinking of goods and services that would have to be foregone, such as a convenient mode (outside of rush hours) of transportation like the automobile.

We also maintain that government planning must play an important role in the ecological struggle. This role evolves not as a replacement for the market system, but because of the imperfections of the market. For if we can rely on the market system, why is it that we cannot go out and buy clean air? The fact is, as we have emphasized over and over, that pollution, in general, is an externality of production. Therefore, it cannot be handled by the market *as we know it*. See, for example, the discussion of recreational uses of water resources in Appendix B.

We, therefore, recommend a strong government role in the fight against pollution and other environmental disturbances. By a strong role, we emphatically do not mean a heavy-handed role. Our contention is that the market system, despite its imperfections, should be treated with as much care and delicacy as possible. So planning moves should be carried out in moderate increments, and benefits from a well-functioning economy should not be sacrificed to wholesale overnight changes brought about by "ecological panic."

If an important role for planning is accepted, it is reasonable to ask whether a superplan for the entire society should be developed. Certainly, such an approach would fit well with the systems nature of our ecological problem. Nevertheless, an examination of the "system" might make us doubt whether this is the way to proceed. For after all, politics is part of the human ecosystem, with as much right to planning consideration as any other facet of human existence, and the political feasibility of superplans in a more or less decentralized society is questionable. Of course, such plans and related analyses of the total economy are of great theoretical interest.

Wassily Leontieff, inventor of the previously discussed economic theoretic model, input-output analysis, has become interested in including pollution in the total production scheme.[1] Recent research has also included pollution costs in other general models of the entire economy, e.g., the "Walras-Cassel general equilibrium model."[2]

But there is also a fundamental social scientific reason for rejecting the idea of superplans in favor of partial approaches to the problem. The theoretical behavior of a system of pure and perfect competition is relatively well understood.[3] In fact, existing markets are not purely competitive, and monopolistic practices, increasing returns to scale (the supermarket's advantage over the corner grocery), and other awkward facts of life are very much a part of the world scene. Welfare economists, then, cannot dream of establishing an optimum society by encouraging the freest possible play of competition and by insuring equity through the manipulation of income. Instead, they must settle for what is called "second best."[4] This "second best" cannot be determined through simple theoretical principles. Experimentation and analysis of results must, therefore, be an integral part of any planning efforts in the real world. So the piecemeal approach[5] to the economic ecological planning process has much to recommend it from both a practical and theoretical point of view.

Does our theoretical ignorance of what the best or "second best" economic situation is mean that we should give up entirely on economic planning measures for environmental disruption? It must be confessed that such an interpretation is possible; but human impatience being what it is, it is likely that people will try anyway to carry out measures that are intuitively sensible, even if theory admits confusion. One might ask what this intuitive feeling means. We think the answer is that it is intuitively obvious that what the mathematicians call "local maxima" exist. By this we mean that small-scale improvements are usually, if not always, possible by predictive means, even if the absolute "second best" is an elusive entity. We know that projects are planned with confidence, and dams are built, highways are paved, and taxes are imposed in confidence that good will result. If only we bear in mind the possible pitfalls and temper piecemeal economic planning with an experimental outlook and a constant comparison of results with policies, there seems to be no reason why we cannot help ourselves to help the environment, even while we lack important economic theoretical knowledge.

That specific planning measures and regulation schemes will help solve environmental problems is, of course, not a new idea. Air pollution control districts in many areas of the United States perform valuable regulatory functions, as do other state and local agencies. The same is perhaps even more true of water quality, with which quantitative economic planning has a relatively long history. An especially interesting example is that of the Ruhr *Genossenschaften* in West Germany.[6] These associations of water users have set up standards of water quality for highly industrialized river basins. These standards are based on water quality being adequate to sustain fish life. An industry that wants to discharge wastes into the stream arranges it with the association. The association calculates the effect of the discharge on the water in relation to the standards set. The industry is then charged a certain fee to compensate the area for the discharge. Interestingly enough, the association will, in some circumstances, pay an industry not to discharge wastes.

If common sense tells you that something is awfully wrong with paying companies not to pollute, you may be reassured to learn that common sense is probably right. Payments for not polluting are fine for General Bull Moose but not necessarily good for the country.[7]

Regardless of disagreements over details, the Ruhr scheme has become an experiment of interest to planners everywhere. The associations represent a good example of a piecemeal solution to environmental disturbance, and they appear, on the face of it, to be effective efforts. They should serve as a stimulus to general environmental planning.

Assembling the Elements

Every environmental plan will be different, but a set of general planning principles is always useful. These principles or elements have been described above. They may be used as a framework for the analysis and decisions involved in planning. These elements are: (1) *criteria* of success or failure; (2) *policies* that are the actions the planner exerts on society; and (3) *tools* for devising policies and for analyzing results. We present below a brief summary of these elements.

Criteria

In quite general terms, the criteria we recommend are quantitative. In a decentralized society, agreement on courses of action can best be served by commonly accepted measures of value. So the dollar value of benefits, including avoided costs, should be compared to the costs of investment in any planned project. For projects entirely in the government sector, such a characterization is entirely sufficient; for those projects involving the private sector, the benefit and cost to the relevant part of society as a whole, as distinct from a particular producer or set of producers, must be measured. Other than that, the analysis for public or private is the same (in principle).

We recommend that an additional criterion, which we discuss in Appendix C, be adopted as a substitute for complete knowledge of the system: the planned modification of the economic production or consumption processes considered must not cause violent changes in the market equilibrium, at least not without prior testing to find out what those changes will lead to.

We can summarize our overall goals or criteria in the following planning rules, which are to hold for any and all planned projects:

1. The ratio of benefit to cost (the "social profit margin") is to be maximized under the following general assumptions:
   a. External economies and diseconomies, such as smog damage costs, are to be explicitly included.
   b. Implicit limits on cost, such as historical budgetary trends, due to political or financial considerations are to be definitely provided for.
2. The limitations of economic knowledge are to be explicitly included. That is, planned manipulations of the economic system are to be relatively small, and actions are to be checked against results at frequent intervals.

Of course, when we apply these principles to selected economic areas of interest, we must allow for variations in practice for the treatment of different problems. For example, it is easy to talk about small changes when one is dealing with a correction to an essentially static situation, such as the effort to decrease emissions from the combustion of fuel oil in electric power generation. It may not be so easy to adhere to our second principle ("small steps") in such projects as the development of a new industrial city in a virgin desert valley, for example.

Policies

Policies that the ecological planner can use are regulation, taxation, and public expenditure. The behavior of simple economic systems under such different policies is examined in a few cases in Appendix C. Generally, the role of taxation, in permitting the greatest flexibility and sensitivity to economic equilibrium, is emphasized. Public expenditure is a major policy variable in planning as it is now carried out, and it will undoubtedly continue to be. Regulation will always be important for hard-core pollution problems. The role of expenditure and regulations is relatively well understood, at least compared to that of compensatory taxation. So we let taxation play a somewhat large part in our discussion of planning.

Tools of Analysis

Some major tools of economic analysis that can be utilized by planners have been discussed in Chapter 8. Regression analysis helps the planner to determine statistical relationships between variables when cause and effect is not known. Market analysis, and especially the determination of price and supply-demand relationships, helps the planner to predict how changing price and cost structures will modify economic output or benefit. Examples of studies using this type of analysis are given in Appendix B.

Modern mathematical techniques for making policy decisions under various money, labor, and other constraints are exceedingly useful in planning. Linear programming and nonlinear programming are especially useful in many types of economic problems. Examples of the use of this type of programming are given in Appendix C. Dynamic programming techniques are widely utilized in examining highly nonlinear systems and especially in optimizing conditions over long periods of time.

Plans: Possibilities and Choices

We have looked at the basic rationale of planning for ecosystem regeneration, as well as the basic elements: criteria, policies, and tools of analysis. Now we might see how an actual government planning agency would go through the process of first selecting problem areas and then deciding how to treat them. In Chapter 12, we will suggest that a specific agency be set up. Here we only talk about a planning agency in the most general sense, to emphasize that governmental powers of decision are involved. We must remember that we are committed to the piecemeal approach to ecological crisis, and so we do not try to set up a superplan. Rather, we try to choose a number of likely areas in which ecosystem trouble appears to be critical. Obviously, it is not enough to know that the trouble is bad in any particular area; one must also be able to do something about it, and so the technical possibilities of dealing with pollution and other environmental diseconomies must be examined.

Survey of the Badlands

We have talked a lot about the various environmental problems and the need to do something about them. But where does one begin? Are the possibly carcinogenic hormones in roast beef the big problem, or is it rather the aluminum cans jettisoned by tourists in wilderness areas? We can see that such different problem areas must be sorted out in some preliminary way before we can even begin the rest of the planning analysis.

The problems to be considered and sifted by the environmental planning agency may be existing, well-known situations like smog or highway litter. But some problem areas may involve more obscure interactions, and the agency may have to first carry out extended investigations of suspected disturbance sources. Still other environmental dangers are those that involve things that have not happened yet, such as a proposal to build a jetport near a coastal shore bird sanctuary. The agency should be given as much power as politically possible to review such projects before the damage is done.

Once a sufficiently (but not excessively) large number of problem areas have been accumulated for study, estimates of problem priority should be made. A good way to sift out various possible ecological problem areas is to make some kind of estimate of the diseconomy (or damages) to the ecosystem. We have discussed some of the difficulties of doing this in Chapter 5, in relation to health and recreation. We suggested there that some guidance is a lot better than none. For the beef and aluminum can comparison we just mentioned, the rough idea would be to first estimate the loss of life and working hours due to the hormone additive in the meat. To do this, we would have to estimate the probability that the hormone does produce cancer, the probable number of people who consume the roast beef, and the resulting total loss of labor hours through death and sickness. For the case of the aluminum cans, we would have to be able to compare a littered area with a nonlittered area. If people from cities close to the littered area are observed to travel farther, spending more money on gas and more time on the trip to go to the nonlittered area, we can reasonably assume that the value of their time in costs can be correlated with the negative value of litter. Then, we can compare the two cases in terms of the money damage done.

Naturally, such estimates are prone to error. But in any sensible planning scheme, some estimate must be made of priorities, or what to do first. So such estimates of diseconomy should be carried out for as many of the environmental problem areas as possible; then, the technical prospects for doing something about the problem should be looked at.

What Can Be Done?

It is evident that it is not enough to recognize a problem. It must also be possible to do something about the problem from a technical or scientific point of view. Of course, if we are all willing to return to a life of hunting and fishing in the forest, jungle, or savannah, as the case may be, we can deal with all environmental problems by eliminating the production or consumption concerned. But people in general don't want to do that. So a great deal of the planning analysis carried out by the agency in this feasibility stage has to be devoted to determining whether some modification of the production process can be carried out to eliminate or reduce the diseconomy, or whether some alternate substitute product may be used that does not pollute the environment. If such modification or substitution is readily available, the particular ecological problem can be cleared for further analysis and economic treatment. If ways to modify the production process within a reasonable length of time or to do away with it entirely by substitution are relatively scarce, the planner should recommend that the problem area be given appropriate priority and financial support in a research and development program. In many cases, a problem may deserve both types of treatment. A good example is that of automobile exhaust. There are modification devices available to reduce the diseconomy under present technology, that is, the "clean air package" engine design and catalytic afterburners. On the other hand, for complete treatment of the problem, a total replacement of the internal combustion engine might be desirable. Because of the time delay factor involved, such a replacement, by an electric automobile or other device, might then be an appropriate field for research.

RESEARCH AND DEVELOPMENT   When severe technical problems stand in the way of getting rid of certain ecological diseconomies, the planning agencies should give support to research studies in related fields. Good research is a notoriously mysterious process, and an exact setting of priorities and goals may be impossible at this point. Still, it may be possible to set guidelines for the amount of financial support that will go into particular problems, on the basis of their priority as determined in the survey of diseconomies. Of course, the probability that necessary research will be taken care of by private companies under the pressure of market forces must also be taken into account. We talked about research on electric automobiles, for example. Considerable research in this area is being done by private firms already. It might be better, then, for government supported research to concentrate in areas such as the development of new types of fertilizer. The widespread use of fertilizers in the United States has been a big factor in destructive eutrophication of lakes and streams. This eutrophication, or overfertilizing of the waters, tends to produce a spectacular growth of algae, which is all right for lovers of aquatic flora, but which kills fish and discourages fishermen and addicts of stream trout sautéed in almond butter. If the government does not underwrite the research in this case, it just might not get done.

Public or Private?

We have arrived at the stage at which the agency has decided that something can be done about a specific problem. Now the question is, how to do it. As far as the public planning agency is concerned, the big choice is between direct government action and manipulation of the private economy. The government can pay for ecological regeneration projects themselves, or they can give loans or subsidies to private firms for the same purposes. On the other hand, the government can regulate or tax the private sector in an effort to modify the production or consumption of goods in an ecologically desirable way.

The decision as to the choice of sector may often be dictated by tradition and custom. It has usually been politically difficult during the past few years for the government to take over parts of the private sector, at least in any overt way. Certainly, when it comes to problems of automobiles, on the one hand, and National Parks on the other, there seems little doubt as to which way to go. The case of electric utilities may be somewhat different. Many utilities are privately owned by regulated and taxed companies, while others are owned and operated by municipalities. The ecological problem of emission from fossil fuel plants might, therefore, be an example of an area in which, for legal reasons, the federal government might have to employ different methods for different plants, depending on the type of ownership.

In general, we tend to favor as much decentralization as possible. This controversial question must remain a subject of political philosophy. What is not controversial is that there will be plenty of work to do both in the public and private sectors.

Government Spending and Loans

Direct expenditures by government play an important role in such fields as water quality, in which there is a long tradition of "public works." These "public works" are variously assigned to the province of the federal, state and local governments. The federal government has often participated in these projects by the funding of pilot plants or by contributing all or part of immediate project funds. In other environmental areas, such as National Parks, the government has a well established monopoly on action. When an ecosystem regeneration project has been assigned to the government sector, it must be compared to other possible government actions. Fortunately, the tradition of cost benefit analysis and quantitative evaluation is fairly well established in government planning, especially in the field of water resources. The analysis of health and recreation benefits is at a more primitive stage of knowledge, but analysis of these factors fits into the general principles of project evaluation as understood at all key levels of government.

We must qualify the above statement in regard to some

proposed new areas of governmental action. One idea, which is somewhat novel, is that of a direct subsidy to nonpolluters, either at the producer or consumer level. Such payments are a feature of the Ruhr Basin scheme and have been proposed at the consumer level for various facets of the solid waste or litter problem. For example, the subsidized purchase of old automobiles by the government has been suggested as a means of solving the abandoned auto problem. We have already criticized such "negative taxes," for an important question comes up that should, ideally, be asked about all government spending: Who pays? We are used to relying on such features of public financing as the progressive income tax to insure a tendency toward equity in the allocation of resources for public use. A comprehensive review of the equity features of established fiscal policies is politically impractical anyway. But we might well question the equity of new public payments for the disposal of old cars, if we are contributing indirectly to monopolistic profits. Monopoly is an accepted part of our way of life, like it or not. But certainly if extra profits are derived as a result of insufficient care in manufacture that reduces the life of the automobile, the source of at least part of the disposal subsidies should be the manufacturer. This question of monopolistic profits comes up also in the taxation proposals we talk about later.

Such new features of the government sector aside, the criteria and policies to be used by the agency in the government sector of the economy are clear. The tools of modern economic analysis are also widely understood. So excessive emphasis on the government qua government planning sector would be the amateur instructing the professional. But we would still like to illustrate here one or two possible uses of analytic planning tools.

A typical water resource case in which the federal government might have an important primary or supporting role to play would be the design of a combined water treatment and sewage plant system for a river basin. For a one-city area, the water coming in must be, in general, treated to bring it up to established water quality. On the other hand, if political or moral constraints are lacking, the only treatment of sewage the city need supply is that which feeds back into the city's own water supply. For a river basin containing several cities the question is, of course, different. Regional planning authorities are concerning themselves more and more with this type of problem. If we look at the details of the situation, we can see that there is always a possible trade off of costs between a city upstream treating its sewage and a city downstream filtering and chlorinating its water supply. A short sketch of the mathematical treatment of such a situation is given in Appendix C. Exact mathematical calculations are extremely useful in deriving what the exact engineering decision should be as far as the building of plants and their operation in various cities in the basin. For any concrete case, the exact answer is hard to see intuitively but can be entirely determined within the limits of the facts available (this particular problem might be usefully treated by the use of nonlinear programming, one of the decision tools we described earlier). So government planners could safely recommend that funding support be given only to that configuration of plants which comes close to satisfying the mathematical optimum, or in other words, that which gives the most for the money.

The government may have to play an important role in the dynamics of economic growth as it affects ecology. Often the development of new industrial areas cannot proceed unless government supplies or approves central support services, such as water supply, roads, utilities, etc. When a new area is opened, it may happen that the first few industries create very little diseconomy in the form of smog, water pollution, and so on. The amount of diseconomy created as time goes on and industries increase in number and scale may grow disproportionately large. In Appendix C, the mathematical form of such an economic development form is briefly sketched. The analytic tool to be used is that of "quadratic programming." A typical example in dollars-and-cents terms might be that industries in the basin might ask for supporting services adequate for a product output of $150 million. The planners might estimate that for this $150 million of output, the amount of diseconomy produced in the form of smog and other kinds of pollution would represent a minus value of $112.5 million. The estimated project net benefit or social product would then be $37.5 million. The government agency would then recommend, in this case, from an examination of the mathematical model, that support facilities be supplied instead only for a total production of $100 million, because this would correspond, under the assumptions described in the Appendix, to a total diseconomy of $60 million, or a net benefit or social profit of $40 million, compared to a net of $37.5 million for the case of greater product output.

The above examples are, of course, artificial, but the kinds of calculation are those ordinarily made in the most sophisticated kind of government planning. This is not to say that there should be a mechanical reliance on mathematical methods or that insignificant differences in calculations are to be relied on; but hard-headed use of actual total projected costs and returns, instead of mere intuitive guesses, are well known techniques in government planning. When we come to the private economy, the situation is different.

The Market Economy and the Government

When the agency planner faces the problem of manipulating the private market sector, he sees that he must depend not only on his own actions, but on the response to his actions by the private entrepreneur. If all of us are conscious, as we should be, of the difficulties facing both the planner in his work and the entrepreneur in maintaining a viable and profitable business while satisfying the moral demands of society,

169

we can see that we need some kind of guidelines for action. A summary statement of guidelines that might be useful to us in the ecological crisis is: *We must influence the private businessman or corporation to help improve our environment, and to bring this about, we want to use for our own social purposes the powerful driving force of the market economy —the desire to make a profit.* This "influence" will often be extended through government policy actions. As policy instruments in this interaction of the government with the private sector, we have, in addition to the loans and subsidies mentioned above, the possibility of legislation (or regulation) and the imposition of taxes. With regulation the government forces the private businessman to do what it wants. In principle, passing a law and enforcing it is a straightforward method of implementation. In practice, it may be that taxation is an important method that we can use in getting the job done. Anyway, the difference between a money fine and a tax is not always that great, and the use of regulation and taxation really should form a continuum of government policy. Also, it appears that we may need every instrument of policy which we can find to deal with the crisis as it faces us today.

REGULATION It requires no great insight to understand how to use a policy of regulation. The legislature passes a law and says that such-and-such is ordained. What is difficult is to decide what the law should say. In general, the problem with regulation has sometimes been that the laws have gone too far in public (or really, bureaucratic) opinion so that enforcement has been lax and producers have gone on creating smog and other diseconomies at will. The other side of the coin is that sophisticated politicians have taken into account both the great power that industry has and the reluctance of the public, when it gets down to cases, to classify the production of certain externalities as a crime. So these politicians, then, have often been able to pass weaker laws that can be enforced. Unfortunately, these laws are often inadequate to the problem. Consider Southern California and its smog problem. Most authorities agree that the public agencies have done an excellent regulatory job under difficult circumstances. Yet no one would be happy with the present state of affairs in the Los Angeles Air Basin.

But there is no question that in some cases regulation is the only answer. The jettisoning of cyanide wastes by metal producers is a case in point. We can assess, at whatever level we wish, the negative health benefits from people dying from cyanide-contaminated water. But, in general, the costs of correcting improper disposal are so modest that the benefit-cost ratio must be effectively infinite. In other words, we can get a lot for a little just by prohibiting the process in the first place. In fact, the whole industrial liquid waste problem may be a good example of something that can be reasonably treated by regulation. Many local government units already have licensing provisions for liquid waste disposal. The number of liquid industrial waste producers in an area is not usually excessively large, and the total volume of waste compared to that of other waste categories (garbage, sewage, etc.) is generally small. The disposal process is often a problem to the producer, anyway, whether it is done efficiently from an ecological viewpoint or not. So regulation here may be able to provide the small amount of coercion needed to get the problem solved.

TAXATION   From a practical political point of view, taxation has one important advantage over regulation. In general, infringement of regulations must be proved in court by the regulating agency. On the other hand, tax assessments (or "effluent charges") can be made and collected by the government agency. If the taxed producer wants to protest, the burden of proof is normally on him. In light of the case backlog in our courts today, this small difference in legal status may loom rather large in practice.

Taxation has many other advantages. We can tax as little as we wish, thereby reducing the possibility of catastrophic disturbances to the economy. Small taxes stand a much better chance of being politically acceptable to business and its lobbyists than prohibitive taxation or regulation. At the least, the taxation accomplishes an indirect environmental purpose; it can serve to compensate us, the public, for the environmental damage caused by polluters. And in competitive sectors of the economy, any amount of taxation will serve to force out marginal producers and will, therefore, decrease the production of products that involve noxious side effects. It is true that many of the most important sectors of our economy are very weakly price-competitive, i.e., monopolistic. In that case, we cannot count on taxes to decrease pollution as much as we might like. But if prices are sufficiently near their monopolistic "ceiling," we can be sure that the monopolist profit will be decreased by the tax. This itself will tend in the long run to encourage the use of alternate, less polluting production processes.

Unfortunately, the monopolist is often able to raise his prices and, therefore, pass along the tax to the consumer. But in that event we are certainly no worse off than before, if the tax revenues are used constructively for the benefit of all of society. These social benefits need not even be connected with the externality, even though an earmarking of pollution taxes to help pollution might have a desirable feedback effect.

We must also remember that even monopolistic prices cannot be raised indefinitely. After a while, people will begin to buy other goods as substitutes for the taxed goods, and the total output of the monopolistically produced goods and their correlated diseconomies will decrease. And, as far as it is politically possible, taxes can be steadily increased until such a price ceiling is reached.

Objections may be made to this type of policy by the producer, in that he may claim it is unfair to tax a process that was thought satisfactory ten or twenty years ago but which is now defined as environmentally disruptive. We believe there is some merit in this argument. As shown in Appendix C, we believe that it is possible to levy taxes so that, when the costs of pollution cannot be shared equitably between the producer and his employees and suppliers, society as a whole will share the cost of the ecological disturbance. This may be thought of as a penalty to society for lack of foresight in not noticing these problems fifty years ago.

Once we decide on a tax policy, we have to determine the amount of taxation. There are at least two very different types of situations here. If an industry is fairly competitive, the goal of taxation should be to recover the damages ("marginal social product," see Appendix C). But since pure competitive industries are very sensitive to cost changes, such a tax goal should be approached gradually, in small spurts, in order to check dislocations and other disagreeable effects. Since there are often cumulative effects in pollution externalities, the final total amount of social tax may be less than originally calculated anyway.

In the ideal competitive world of Adam Smith, the use of such taxation would provide the bonus of an "optimal use of economic resources." This, then, would be an additional advantage of taxation over regulation. But in the world described, for example, by John Kenneth Galbraith's *American Capitalism,* we must be content with more modest goals.

For there are many industries in which not much real price competition takes place. In these monopolistic or oligopolistic cases, there is often excess profit available that can be tapped by taxation. But that does not mean that taxes can routinely be set so as to drain off all the producer's surplus of the monopolist. For political reasons alone, small but growing tax assessments are probably best. On the other hand, the final limit on taxation, while it can be set at the "fair share of dam-

173

ages," need not be limited to this amount by any ethical or economic imperative.

We do not believe there is any final answer available right now to the general tax assessment question. For the sake of discussion, we have proposed a formula based on comparing "fair shares" (monopolists vs. the rest of us), as they would be first, with better production processes (less bad externalities) and then with the existing process and compensatory taxes. The monopolist can then be asked to pay (as an eventual goal) his share of the damages plus the excess profit which he derives from using poorer (more diseconomy-producing) processes.

As an example, consider the problem of taxing those industries in a certain air basin that produce an amount of sulfur dioxide as a byproduct of their operations. We assume that we can cost-out the SO_2 diseconomy (from deaths and sickness due to bronchitis and emphysema, cf. Chapter 5) at $10 million to $20 million per year, and the number of units produced is assumed to be 150 million a year. The mean cost of the diseconomy per unit is then 10¢, taking $15 million as an average figure for the diseconomy. If the industries are competitive, a reasonable plan, neglecting tax collection costs, would be to set a tax of 1¢ per unit initially, and keep increasing it until the industries pay for the "damages" 10¢/unit). This scheme assumes that the tax cost can be "passed down" or "shifted forward" by the producer. If the industries are monopolistic, we might want to begin the tax plan schedule the same way, but eventually we should like to do a cost analysis of industries and set the maximum tax in accordance with the available monopolistic profits and with alternate methods of production. If the units sell for $1, and excess profits are estimated at 10¢, then the final tax might depend on the possibility of alternate (better) product processes. If alternate processes are very cheap, the total tax to the monopolist (and his employees and factors) would be again 10¢ per unit. The excess profits are due to the market alone, not to saving money by bad production processes. If the better process would cost 5¢ per unit more, then one could attribute the monopolistic profit as partially due to avoiding these extra costs. The final tax on the monopolist (and his supplies and workers) would then be greater than before. From the formula in equation 19 in Appendix C, we calculate it at 14¢. Other lines of reasoning are also possible for determining the exact amount.

Naturally, the above example is very simplified. For example, the role of price increase must be taken into account, and the spreading of the tax among suppliers and labor must be calculated. These difficult questions can be treated, at least in principle (see Appendix C, equations 11–16). It may well be objected that costs, profit margins, etc., are impossible to determine, but we think the truth will out. Given a gradually increasing tax program, the actual behavior of producers will reflect the real (perceived) cost and profit structure. Errors in estimates can then be checked against what actually happens.

This uncertainty in measurement can be related to another problem: what about the firms that are neither perfectly competitive nor perfectly monopolistic? Certainly, most firms must fall into some intermediate category of imperfect oligopoly or monopoly or monopolistic competiton. In a theoretical sense, such firms would have to be analyzed to determine whether or not excess profits are available to help pay environmental taxes. Even if a firm is identified as a "monopolistic competitor," however, it may not be possible to determine whether profits are normal or not. If it is difficult for new firms to invade the industry, it is probable that excess profits will exist.[8] Otherwise, the situation, as far as profit margins are concerned, may be the same as in perfect competition.

At any rate, in practice, as we have pointed out, it may be impractical to determine the profit structure of individual firms, much less that of entire industries. This is especially true in the context described in Galbraith's *New Industrial State*, where profit levels are *maintained*, rather than *maximized*. Under these circumstances, it is plausible that unnecessary, but tax-deductible "expenses" take the place of "profits" on the corporate ledgers.

So the gradualist scheme of ecotax imposition we have suggested may be necessary, not only because sudden dislocations should be avoided, but also because of our ignorance about price and profit levels. In practical terms, then, we should be prepared to test techniques and examine results as we go along. The principle of environmental taxation appears plausible, the interaction of taxation with the market must and should be investigated over the course of time.

It would be not only immodest, but foolish for us to claim that taxing procedures such as that in the example above are either new in concept or infallible in detail. Since something must be done, the details are not all-important, but the awakening of informed, intelligent concern to the possibilities of such schemes is.

We have mentioned before the possibility of placing a tax on consumers instead of on producers. There is no reason why not, from an environmental point of view, and many reasons why, when we consider the political facts of life. Indeed, when monopolists are able to raise prices to cover taxes, the tax might as well be on the consumer in the first place. For competitive industries, the precise difference in result depends on the details of industry supply and demand functions (see Appendix A), but the general result is often the same. One mundane answer to the question probably lies in the political sphere. Consumer taxes are visible and, therefore, unpopular at general election time. On the other hand, producer taxes are guaranteed to stir up hornets' nests of agitated industry lobbyists. But we dare venture no further into this murky "polluted" realm of affairs.

There are several special areas in which consumer taxes

might be of special importance. One is the consumption of durable consumer goods. A special case of great interest is the possibility of taxing automobile use for environmental purposes (in the previous example the auto maker would be taxed instead). This consumer tax would, in effect, be a levy on the average production of pollution by car users. But if we tax the users per mile (by additional gasoline or odometer taxes), the rate of use and, therefore, of pollution would be lowered. The advantage here would depend on an indivisibility of the market—the overwhelming consumer need to have an automobile for some occasions. The "use-type" tax might be more effective (at the same revenue level) in getting around this difficulty. Of course, use taxes do tend to bear most heavily on the poor. A supertax on gasoline might well have the effect of discriminating against leisure at the lower economic levels. So if we are to maintain even the present level of equity, we should adjust other taxes to compensate for income discrepancies.

There are still other cases in which there is a real economic difference between producer and consumer taxes, in addition to possible vagaries of the price structure. For example, the geographical location of the firm or consumer may be important. In many instances, the use of fertilizer may be viewed as nothing but a benefit for the farmer and for the nation: agricultural yield is raised and there is more food to eat for all. But in some regions excessive runoff of fertilizers has caused severe damage to our watersheds. We mentioned this problem before in a research context. It can also be considered in this taxation framework.

What should the government choose as a taxing policy appropriate to the fertilizer-eutrophication problem? An obvious answer is to tax the use of fertilizer in those regions where it is a problem. The farmer certainly has a right to try to increase his crop yield as much as he can. In doing that he is helping himself as well as helping the rest of us. But in help-

177

ing to destroy life in the streams and lakes around him, he is hurting both himself and us, and for that byproduct of his increased farm production, he should be made to pay a compensatory tax. It, therefore, appears that the government should feel free to use maximum flexibility in taxing, as well as in regulation.

It might be a good idea to also mention the population (or "popullution") problem here, since the number of people affects both production and consumption, and so plays a critical role in determining environmental disruption. Would it not be possible to tax production of children?

The control of population has been widely touted as *the* key to the ecological problem.[9] In the long run, of course, population must be controlled by man, or, if not, it will be controlled by nature. But in the long run, too, the resource recycling problem must be solved, regardless of population levels. A world of 500 million can waste resources for a longer period of time than a world of 5 (or 10 or 20) billion people. Eventually the judgment day will come, and the general utilization of governmental powers to direct private production and consumption is inevitable, if we are to survive, no matter how many people there are on earth (or Mars, Phobos, etc.).

So we should like to emphasize the planning controls we think both unavoidable in the long run and politically practical in the present. Population control, we think, is both desirable and inevitable, but it is, despite large amounts of wishful thinking, extremely controversial at present. This controversy is perhaps more tied to love than religion, and children are a highly prized consumer good in many, if not all, human cultures. We, in highly industrialized countries, are perhaps fortunate that we do not have to breed children to relieve labor shortages in the family economic unit.

Certainly, the taxation of childbearing is administratively feasible, just as the subsidy of families (in Canada, for example) has been a practical policy. If the critics of "popullution" can solve the political problem involved, more power to them, but the zero population growth movements should not be allowed to elbow aside the other aspects of the ecostabilization problem.

In summary, the policies to be used will depend on the criteria involved, as well as the economic background. The exact policy values can be determined by using the mathematical tools of economic analysis. Anything is possible if we take everything into account, including irrational factors, but humility in the face of our gaps in knowledge of cause and effect must dictate that we proceed carefully.

This sketch of planning procedures should give some idea of how to proceed in playing the market economic game for ecological purposes. Plans can be made for air pollution in a given air basin, for water quality in a river estuary, and for the problem of littering the countryside with aluminum cans and plastic bottles. As many piecemeal plans can be undertaken as are politically feasible. Each plan should proceed cautiously with well-defined initial limits on taxes to be imposed, with an overall emphasis on rationality, and a consciousness that we must not throw the economic baby out with the polluted bath.

Budget and Politics

We have been assuming all along that various ecosystem regeneration projects can be selected, analyzed according to the technological feasibility of solutions, and assigned to subsidized research or, alternately, to action in the form of a policy project to correct the situation. But the realist in government will point out a criticial missing element. What about the necessary government money? The amount of an initial budget to operate a suitable planning agency will be fixed by the political process. And the process is truly political in that the allocation of government funds must be stimulated by people in government and outside government, who are convinced that something solid, down-to-earth, and comprehensive must be done about environmental disruption. That initial process really lies outside rational planning, but we may ask what the role of budget is once governmental ecological planning gets under way. In other words, what about next year's budget? The essential element, of course, is to show that the present year's budget has been spent efficiently. For that purpose, the assessment of diseconomies, the analysis of the state of technology, and the determination of the cost of environmental correction that will have been made for planning purposes will be of great use. As we have stressed, although evaluation techniques are not down to the level of an exact science, a national ordering of priorities will be helpful in two ways. We can show that it is wise to build water treatment plants instead of sewage plants (or vice versa) this year; at the same time, we can point out that a program for the taxation of plastic containers, although we did not have enough money to administer it this year, is still of the highest priority and deserves funding next year. The purpose of talking about ecological planning is not that the people should be encouraged just to "go out and spend some more money," and the public should be satisfied that they are getting their money's worth in the ecological fight.

To Err Is Human

It might seem fatuous to emphasize that people can make mistakes; however, when it comes to planning models, there is a whole separate subdiscipline called "sensitivity analysis" which deals with nothing more than seeing how important a mistake might be. It is really not to the point to go into this

subject at length here. But it may be reassuring to some to realize that planners do have fallibility (their own as well as
that of others) in mind when they carry out their mathematical analysis. Most facts in economics are merely things that
are probably so, not sure things. If one is conscious of this,
one can make everybody feel better about a plan if it can be
shown that the resultant policy (taxation, regulation, etc.) is
still the best policy, even if all the data are not quite right.
This is good principle and a simple concept but one that is
not always observed—it should be.

Feedback

Implicit in what we have been saying about caution and fru-
gality in public expenditure or taxation policies is that we
must make provisions for constant reappraisals of the plan.
From these reappraisals, we can make changes in the plans,
increasing or decreasing the amount of taxation, shifting public expenditures, etc., and, in general, steadily improving our
picture of how the campaign against environmental disruption is progressing. Generally, there will be many feedback
mechanisms involved. Often we will have to consider internal
feedback, such as the failure of effects to go along in a
straight line. Unfortunately, many mathematical planning
tools explicitly assume this straight line behavior. Frequently,
it is either computationally impossible to use more complex
mathematical forms, or the quality of the data is such that a
complicated theoretical form is not justified. But after the
plan has been in operation for a while we can readily adjust
it to what is actually going on in the world.

Many possibilities come to mind. A simple example comes
about in the development of recreational water resources.
The amount of boating typically does not depend on just the
amount of money we spend each year on boating facilities.

181

Usually, the provision of even minimal facilities causes a great spurt in boating activities. In this case, our plan is faced with an internal "nonlinearity" of behavior, and appropriate corrections must be made.

Such feedback mechanisms can also come into play in a more complicated, external way. One production process affects another. A good example of this would be a possible shift from gasoline engines to electric automobiles to reduce the amount of smog. Remembering the concern voiced nowadays about the pollutant effects of fossil fuel electric plants, we would be forced to realize that this would mean a decrease in the amount of PAN, an eye-irritating ingredient, but an increase in the amount of sulfur dioxide, a different type of tissue irritant.

So it is obvious that odd things can happen, and we lose something by not having one grand superplan to take care of everything in the future. But given the world as it is, the cautious approach seems better suited to our ecological purposes.

We can talk about the restricted kind of feedback from present production processes and actions taken against diseconomies as they affect the planning process. We can also look at the way in which the changing world should feed the goals of the planner. It has been pointed out that the world is changing from a "cowboy" to a "spaceship" economy,[10] and that production without regard for waste is on the way to being replaced by conservation without regard to production. The point at which this changeover should take place is again a political decision and will depend on public attitudes and the particular human value systems existing at the time. The long-range tendency is toward a society in which the conservation of the materials inventory as the all-important factor seems to be indisputable. In the present value system of economic planners, the amount of resources available to carry out production is taken as restricted, while the amount of production is a variable factor, which we try to make as large as possible. In the "spaceship" world of the future, we will instead demand a certain subsistence level of production and consumption, presumably one that will be adequate to satisfy our needs as we understand them today, while the amount of precious elements in our earth ecosystem will be a variable. And we shall try to make this variable as large as possible in order to make our tenure on earth as secure as it can reasonably be. An example of how mathematical programming can be used for this kind of "spaceship" economy is given in Appendix C. Regardless of specific calculations, it is evident that in order to plan our economy along these lines in the future, economists will have their work cut out for them. Data on stores of resources are as yet rather limited, so students of the future will have plenty of material for research.

The "spaceship" point of view is merely an example of the kind of physical-scientific fact or human value judgment that may change planning goals as we go along. Naturally, any planning organization must be organized to keep abreast of such developments and to incorporate them into the actions of future governments in the economic sphere.

Enforcement and Reinforcement

The interaction of production processes and the stimulation of consumption by production can, as we have just seen, affect the outlook for ecological economic planning, but we must not forget that politics and propaganda also come into play. We have talked at length about the necessity for using economic means to solve worldly problems. Inevitably, we must make an important exception for a basic structure that underlies all plans: politics. If taxation schemes are undertaken, they will be undertaken by political bodies. As we examine the feedback in our planning models from time to time, we can recommend changes in policy, a review of taxation, a shifting of federal funds from water pollution research to solid waste investigations, etc., but, fundamentally, the economic models must serve to promote a common basis of understanding among at least a section of government officials and people. Ultimately, some citizens will have to influence other citizens. Taxation plans must not only be rational and understood, but they must be passed as actual laws. So nothing in this book is to be taken as a slur at the efforts of environmentalists to organize public opinion for the relief of the ecological problem. More propaganda, more speeches, and more concern by enlightened citizens will always be needed, if the problem of environmental disruption is to be treated intelligently.

10

The metaeconomic planning methods that we have described in general terms may be useful in modifying the future shape of the world so that it corresponds as much as possible to our perceived preferences as human beings. We have concentrated on general procedures because we think that the general feasibility of the idea should be emphasized. This general feasibility also should not be obscured by the inevitable imperfections of specialized plans. Special plans are essential to concrete results, and the formulation of such plans may be carried out as discussed in Chapter 9. The precise shape of the world to come will depend on how these plans shape up as part of the larger matrix of human actions and environmental consequences. Naturally, we are all curious to know what the future course of the grand ecological system will be. A few general tendencies of the world of the future are discussed in this chapter and in the next one. In this chapter, some successes, or various alternative equilibrium ecosystems, are looked at. Failures, or unstable ecosystems, are examined in the next chapter.

185

Is There a Best?

First of all, let us see how many "optimum" worlds there are. Ideally, there might be only one best world, the one we should strive toward by the application of rational planning methods. Which world this is, is not necessarily knowable beforehand. The best world will arise from the best treatment of its component parts; but usually we know at most only something about some of the components of this complex world. A tenuous analogy in microeconomics might involve the variety store manager who tries to get the best profit margin he can from each individual item, without necessarily knowing beforehand what his total profit will be. The description of a metaeconomically optimized world can only be obtained by going through all actual calculations for such phenomena as various details of combustion processes, the sale of plastic dishpans, and the recycling of agricultural water into industrial use, to cite a few of the myriad problem and planning decision areas. With this accomplished, however, a world picture could be painted of the optimal ecosystem that could modestly be called a planned utopia of the future.

Of course, we are all rightly suspicious of the word utopia, and there seems to be little possibility that one will actually face such a phenomenon. In the first place, we must realize that the best may not be very good, the optimal system being only the best possible under the circumstances. The optimal ecosystem for earth four billion years ago was a very hot conglomeration of relatively unstructured minerals. This might not satisfy us nowadays!

More serious difficulties in the utopian picture, though, arise from the fact that any plan is only as good as the data that go into it, and the data for many concepts, especially those dealing with esthetics and unformed preferences, are very shaky indeed. The final shape of the world, if it follows successful metaeconomic planning lines, would be actually a "band" of possible ecosystems, to borrow a term from solid state physics. Even more vexing than the planning problem, pure and simple, is the related difficulty of giving adequate theoretical and practical estimates for metaeconomic criteria. All of the many values treated in the metaeconomics of ecology have been discussed both in terms of welfare economics and in various engineering applications, and though the principle of equivalent prices (for nonmarketable items) is a way of estimating and checking criteria values, the chance for error is large. Errors also creep into the execution of any plan. The political process, in this sense, may be held to be an example of error, i.e., if you don't agree with me, the eco-planner, you must be wrong. In both democracies and autocratic regimes, differences of opinion affect even the most highly organized network for relaying decisions.

Now is it fair to ask if there is anything left of our recommendations for planning? The answer is a decided yes. Although a respect and understanding of statistics has not yet penetrated as far as the cost-of-living averages (nor to the point scoring system in boxing matches), the fact remains that we are well able to tell when an answer will be right 9 times out of 10 or 99 times out of 100. This kind of result, needless to say, is so superior to the kind ordinarily involved in the conventional wisdom approach to decisions in everyday life that one need not be apologetic for metaeconomic planning models. To be sure, uncertainties will exist. All uncertainties, both in data and theoretical formulation, will be estimated as accurately as possible and incorporated into a statistical framework. The net result will be, if not a best world, a group of planned worlds that are, hopefully, somewhat better than what we could do otherwise.

We will take a brief look at some of the general ways in which such a planned world may develop.

Our Alternate World-Lines

The general appearance of the world in the future will vary according to the decisions we make within the next few years and decades about our relationship with the total ecosystem. In the physics of general relativity, in which time is considered at a formal level in the same way as spatial dimensions, every particle in the universe has a path through both space and time. This path is called a "world-line." We extend this nomenclature by analogy to think of time as a dimension in a frame of reference in which the various aspects of quality of life on the earth make up the other dimensions. We may, then, think of the different possible worlds as lines diverging into the future from a central point, which is our present society on earth.

The first kind of world which we might consider is a world very much like the present one. In a way, such a civilization would be perhaps the most straightforward extrapolation from our present state. The naturalness of such a familiar earth may be more apparent than real. We will immediately perceive that the perpetuation of a society similar in nature to our own depends on very particular circumstances, even though the face of the world will be changed by technological advances, such as the introduction of electric cars instead of gasoline powered cars, to mention a trivial instance. Even allowing for technological change, it is evident that the real determinant in preserving the world in its present state is the control of population. If population continues to grow at the present rate, the much commented on urban sprawl in New York and Los Angeles, to say nothing of Tokyo, is merely a mild taste of the future. Most likely, population control will be achieved, at least in the far future. Religious objection is probably overpublicized, since it appears, from the Latin American experience, that the effectiveness of religious discipline in the present day is rather limited. Cultural reluctance, as is probably the case in India, is a more serious problem. The history of the past hundred years is, in a sense, the story of the triumph of acculturation, but the drives to bear children may still have strong resistive powers.

If population is indeed leveled off, will the world as is continue as a viable form? Naturally, the problems of generalized pollution and inequitable distribution, i.e., poverty, would still be difficult ones, but they could surely be solved eventually if the necessary sacrifices are made. So there seems to be no conclusive reason why our present pattern of cities, suburbs, farms and rural communities cannot be modified in such a way as to make it a viable scheme for the human part of the world ecosystem. What we know now as our world may not seem like heaven to us, but a little reflection will show that it is far from the possible hells that await us if we are not more careful about the way we do business with our environment.

What if the population explosion cannot be contained in the relatively near future? Presumably at some point in time population control must enter, even if it means colonization of other planets or solar systems. A near-term population explosion could well lead to what we may call a "cement civilization." The present urban sprawl could become a cancerous growth over the whole earth. Of course, the configuration of the supermetropolis would have to include farm factories, just as cities now include industrial plants, or assuming the oceans are not covered by a type of houseboat community, food would have to be derived directly from the sea. It is even conceivable that some way will be found to alter human genetics so that people could be bred to have cells containing chlorophyll, so they could absorb sunlight directly, like green plants, and make their own starches and sugars. Even neglecting such extreme hypotheses, we see that a world of one supercity is indeed possible. While the proper preservation of wild spaces, for example, might be reduced to the efforts to save a few large parks, there is no reason why life on such an earth, though seeming somewhat odd to us of present-day society, could not be a perfectly feasible *Brave New World*.

It is also probably true that we have not yet plumbed all the possibilities for ameliorating the effects of congestion. A closer attention to urban esthetics would naturally be essential if eye pollution is not to become overwhelming. Also, an extension of some of the tricks of concealment of the landscape gardener may help. For instance, in the California mountain resort of Lake Arrowhead, newer subdivisions are based on a 3/4 acre minimum lot size. At close range, though there are trees and pine needles in every yard, the effect of wilderness is quite restricted by the sight of the surrounding houses. But looking across at the lots and houses from the hill opposite, the "resort sprawl" is quite effectively disguised, and the hill appears to be an unbroken forest of yellow and sugar pine, black oak, white fir, and incense cedar. Such large-scale cosmetic effects may be of great importance in a future world. It goes without saying that pollution and distribution problems will necessarily have to be solved as a precondition of the "face in the crowd" civilization. It is probable also that some type of continuous group encounter experience will be a necessary replacement for the wars, sporting events, and tea parties of the present world, and euthanasia for the elderly might replace birth control procedures, if a truly unrestrained birth rate continues to be public policy. Of course, in such a world senility may be defined as beginning at age twenty-six!

Zero population growth and unrestrained birth rates are not the only two possible outcomes of the present population crisis. There seems no good a priori reason why the population of the world should not actually be considerably reduced from what it is now, although again cultural or psychological evidence can be adduced to the contrary. An interesting consequence of such a population reduction is that it would make possible another different type of world-line for the earth of the future. One might speak of a reruralization of the world with cities replaced by networks of small towns, farms, or communes. The popularity of such an approach has been attested recently by movements among certain tuned-in sectors of society for a return to the "open land." Many such communal or collective farms have been formed within the past few years. Whether this approach can have much success or not in the present state of industrialization and technology is another question. It appears that many communes studied actually depend on outside earnings, or on remittances from the "straight" world, in the form of welfare or unemployment payments.[1] But there is certainly no bar to the success of such communal efforts if the expected level of product consumption can be uniformly lowered throughout the world. Such a scheme has at least the advantage that the problems of pollution should be much easier to treat than in our present highly industrialized civilization, *ceteris paribus*. Here again, though, one must take into account cultural effects, since, for example, different notions of sanitation among the new generation and consequent local pollution of water supplies have occasionally led to a growth in intestinal flora to levels perhaps more characteristic of such places as Managua, Nicaragua than of the surrounding straight communities. So a reorientation of some ingrained psychological inhibitions plus a reallocation of health resources may be necessary in such a future communal world; but, of course, the history of man is a study in reorientation.

We have mentioned only three very general life patterns of the future for a viable ecosystem. These patterns are critically dependent on the "capita" part of per capita income or per capita welfare, otherwise known as the population problem. Naturally, there are many choices or world-lines available, including combinations of those mentioned and undoubtedly other configurations. But it is well to keep in mind that we have some choice of type of future world, other than a perpetuation of the present, and that our attitude toward the future world must take into account the costs and benefits of the life quality in such future civilizations.

The Depletion of Vital Resource Banks

We have previously alluded to the long-term problem for continuance of any civilization that depends on a high degree of industrialization. We must develop new means of conserving or recycling vital material, such as coal, oil, iron, sulfur, etc. It goes without saying that the recycling problem of oxygen and water must be solved if we are to survive at all. Methods of accomplishing this basic recycling (through careful treatment of water supplies and preservation of oxygen-generating plant life) are known to us. For the recovery of minerals, however, we have to face the fact that industrial processes often greatly alter the chemical or physical form of the elements used. Recent arguments about air pollution due to use of gasoline combustion stress the necessity for preserving petroleum supplies as a source of vital hydrocarbons for chemical use; the same argument can be applied to coal. When we discussed the shape of worlds to come, we did not mention that the complex problem of mineral recycling may well change our society in ways which are hard to foresee. It also seems obvious that we must account for this factor in figuring out the economics of waste disposal. Recent reports estimate that, for example, all of the solid waste in Los Angeles County could easily be taken care of during the next fifty years by sanitary landfills. In 5,000 years, is it not possible that people will be engaged in a new industry, that of digging up sanitary landfills in order to process the metals and other materials they contain? An obvious solution here is the use of advanced machines to sort and reprocess many of the elements of the wastes. These elements are not economical at the present time, but are they in the longer view? Society has a stake in such questions and the full force of metaeconomic planning analysis should be applied to this question.

Ultimate Hope?

Some of us remember wearing buttons from the 1939–1940 New York World's Fair that said, "I have seen the future." We never thought then to doubt what the future would be. The future was a continuous stream of streamlined General Motors cars gliding peacefully along elevated freeways. The outlook may be less certain now. At any rate, in this chapter we have tried to look at some of the future world-lines of our earth. What kind of world we have depends on what kind of plans we make now. If we do not plan at all, we may have no world. Some consequences of negligent planning are examined in the next chapter.

11

PATHOLOGICAL POSSIBILITIES OR SICK WORLDS OF THE FUTURE

Increases in demand over time for more and more goods and services, and the efforts of industry to respond to these desires, supported by an exponential explosion of technology, have made the ecological problem especially evident today. But what of tomorrow? *Ceteris paribus,* societies can count on one hell of an ecological mess. So tomorrow is a function of today; the evidence of today can support meaningful suppositions of tomorrow. In this chapter, we review a few, and only a few samples of pollutive events, and we then speculate in narrative form on a possible man-made environmental future. Dramatics aside, man ignores the consequences of his economic actions, and nature does not. That is all.

A New Jupiter

The hazards of big-city pollution, where auto exhausts, incinerators, and electric plants pour out soot, dust, and poison gases, are pretty well known.[1] Certainly, the word smog is not foreign to most people's vocabulary, both here and abroad. The 224 ton granite obelisk known as Cleopatra's Needle, carved in 1600 B.C. and presented to New York City in 1882 by the Khedive of Egypt, serves as evidence of what happens in a transition from the clean air of the past to a quasi-Jovian atmosphere of the future. The obelisk maker of Egypt cut hieroglyphic characters into all four of its sides, and the ancient writing was still plainly visible when it was brought to New York. Today, however, these markings are only on two sides; the markings on the south and west sides, which face prevailing winds and concentrations of air pollution, have been entirely obliterated. Several inches of granite have been literally eaten off the obelisk by the chemicals in the air. Ninety years in New York has done more damage to this untended monument than 3,500 years in Egypt. New York's polluted air can eat rock; it can also hurt people.

The London smog of December 1952 and the Donora attack in the late 1940s are often cited as two of the worst air pollution disasters in medical history. On Friday, 5 December 1952, a temperature inversion over the metropolitan area of London resulted in an accumulation of such high concentrations of air pollutants that approximately 4,000 people in the greater London area died as a result of the smog. On Tuesday morning, 26 October 1948, a dense fog began to envelop Donora, Pennsylvania, and again, as a result of a temperature inversion, it remained in the town for five days. When the fog lifted, 20 people had died and approximately 6,000, or nearly 43 percent of the population in the Donora area were suffering from varying degrees of illness. Neither of the pollution disasters was due to accident. Temperature inversions are normal climatological events. In simple terms, the disasters occurred as a direct result of growing air pollution.

The pathological possibility of an air pollution disaster now threatens such huge metropolitan areas as New York City, Washington, D.C., Philadelphia, and Los Angeles, where immense populations live in, if not blissful, at least resigned oblivion and apathy. The event, whenever it occurs, will probably result in several thousands dead, and hundreds of thousands ill, and it may or may not arouse a sufficiently strong public reaction.

Should we continue along the same path we can expect markedly increased air pollution in the next quarter century. In man's continued use of fuels, he will, over time, add to the concentration of particles that settle on the earth. Present chemical processes, if continued, will result in more waste gases or fumes in the air; if incineration is continued as a method of disposal of increasing amounts of household and solid waste, it will contribute enormous quantities of particles and noxious gases; the "transportation explosion," which has yet to hit the lesser developed countries, will result in substantial increases in the use of fossil fuel, and, because of the usually incomplete combustion of this fuel, an additional amount of toxic substances will be continually added to the air. The complexity of the problem becomes more pathological when combined with photochemical reactions in the atmosphere and contamination of the air from radioactive fallout.

The removal of these air contaminants is hard to imagine, except within enclosed spaces, such as buildings. Whatever your vision may be of cities of 100 or more years hence, they probably will not be roofed-over metropolises, where the air is purified as it is brought in from the outside. It is much more likely that there will be cities whose inhabitants wear gas masks to purify the air as it is breathed. At least, the trend is in this direction.

Surprisingly enough, the solution to the air pollution problem can be quickly reduced to the application of various and sundry economic policies to prevent or minimize undesirable contamination at its source. But for hundreds of reasons (political?), air pollution will remain virtually untouched, and it will remain so until the public becomes outraged at the state of its neo-Jovian environment.

What will happen if nothing is done? We may picture the scene in future years, in a world whose primary social sickness is runaway air pollution. Let's imagine that in this sick world of the future civilization depends more than ever on the automobile. Control of exhaust, fortunately, has been at least partially effective. Even so, the amount of photochemical smog grows, year by year. Part of the problem has been the steady growth of population, requiring that more and more automobiles be produced every year. The introduction of miniature sidewalk motor scooters has helped solve the critical short-range traffic problem but has made the pollution problem much worse. Another factor has been the great growth of auto traffic in the formerly underdeveloped parts of the world. Recent smog attacks in Entebbe and Ouagadougou have been especially severe. The incidence of sickness from air pollution has been particularly pronounced in newer nations. These countries usually lack the complex weather forecasting services that protect Megalopolis East in the United States from severe attacks during temperature inversions. Oxygen packs, too, are less widely used in the less affluent nations, and organized screening of urban residents for asthma, bronchitis, and other lung ailments is done only in the United States, Western Europe, and the Soviet Union. Officials have estimated that such screenings in Megalopolis East alone, since their introduction two years ago, have saved 50,000 lives from sulfur dioxide attacks. There have been complaints from those who fail the screening tests about the inconvenience of relocation in rural areas. In the first place, the available rural areas have greatly shrunk in countries such as the United States, since the areas of Megalopolis West, South, North, Middle, and East take up most of the available land. It is true that there are "rerualized" areas, such as the Los Angeles Basin, which was abandoned ten years ago as officially "unbreathworthy", its inhabitants relocated in the Colorado River Basin. But such areas obviously cannot be used for the resettlement of the "breath-unfit."

The cost of resettlement, as well as subsidization of oxygen emergency and portable breathing packs has been allocated, in major part, to government, and has been financed through Social Security by an additional payroll tax of 2.5 percent (which is just 10 percent of the extant levy). So there is little public sympathy, from the urban taxpayer, with the complaints of the resettled population.

There has been some criticism of the government's failure to take action on the growing incidence of a new type of injury to the cornea of the eye. This phenomenon is so recent that scientists are still speculating on the cause. It is believed by many that an unknown reaction between highly concentrated hydrocarbons is the cause and that it is closely related to the simple eye irritations that have been observed ever since 1945. But research is continuing.

The travel business is booming. There are popular five day tours from Europe and America to Baffin Island and Little America for blue-sky-show exhibitions. Tahiti was also popular for sky and cloud enthusiasts, until recently, since the viewing during the last two summers was marred by occasional banks of brownish-gray clouds, undoubtedly blown in from the industrial regions of New Guinea.

Economists are naturally concerned with the high costs of atmospheric management, but the high level of the GWP (Gross World Product) achieved makes experts reluctant to disturb the course of the economy. Planners in newly-developed countries have been especially opposed to changing the operating modes of their burgeoning new industries. Politicians remember too well the riots of workers at Ulan Bator when the Mongolian Council of People's Commissars, at the urging of a Nobel Prize–winning chemist, proposed the closing of a new auto plant and a general cutback of pollution-causing industry.

The status quo in air quality will probably prevail for the near future. Eventually, it is hoped that the replacement of internal combustion engines by electric motors and the conversion of generating plants to nuclear energy will catch up to the pace of current economic expansion, and the supply of fossil fuels will at last become exhausted, even though recent oil finds in the Indian Ocean and in the Kamchatka Peninsula have given petroleum production a new lease on life. Most worrisome, perhaps, has been the indication that a new atmospheric phenomenon is occurring. For reasons not yet properly understood, clouds of pollutants occasionally are wafted up into the stratosphere, collect there for a period of time, and then finally deposit in locations far removed from the original source. A team of physical scientists is now hard at work on the problem on a $150,000 grant from the World Science Foundation. It is hoped that progress on the phenomenon will be made soon, before serious disasters occur.

The world we have just imagined is fantasy. At least, today it is. Let us hope that a tomorrow of this kind never comes.

Water of a Sort

Lake Erie is gradually turning into a new Dead Sea. The waters near the shore are heavily polluted by sewage and wastes. More than a billion gallons of sewage are discharged daily into the ocean along the coast of California. The great estuaries of the East Coast are now so heavily polluted that in many places only a few limited strips of coastline near metropolitan areas are available for bathing, and even these recreational areas are closed from time to time because of high bacteria count.

Pollutants filter downward to the ground water reserves of an area and disappear temporarily from the eyes and nose and the chemical senses of man. When they enter the underwall of the water table, the pollutants participate in an obscure, often unpredictable system of water dynamics. For

199

ground water does not stand still, and its speed of movement varies in different places, often dropping to a flow of only one or two feet a day. Heavily polluted ground water may not appear in a well aquifer for months or even years after the water has become irreversibly contaminated. It is like a cancer that is silently spreading deadly cells to vital organs of the body without causing discomfort or pain. The first evidence of damage may appear only long after it can no longer be arrested—in the case of ground water pollution this may be years later (shades of a *Silent Spring* [2]). This is not all; eventually, ground water appears as surface water, rising from the earth's depths as springs, streams, or the countless unnoticed small sources that nourish our lakes and rivers. Once these water sources are spoiled, they can hardly be closed down like the sewage pipes and drains of a municipality. So lasting damage to the nation's ground water resources reaches to the very basis of our water supply; indeed, it endangers the future of man.

The water cycle is really a wondrous thing.[3] Every day, the sun evaporates something like 4,300 billion gallons of rain water. About two-thirds of this is changed into vapor by the sun's heat and the life processes of plants. The remaining third of this precipitation reaches rivers and lakes as runoff or becomes ground water. In the United States, a great deal of the precipitation falls on densely populated cities, where heavy rains help to promote rather than diminish water pollution. This is easy to understand when one considers that some of our largest cities also have the oldest sewer systems. When these sewer systems were developed, they were designed to collect storm water and sanitary sewage in a single system, and thereby saving on cost and the installation problem of two separate systems. At the time it was probably good economics, but today these combined sewer systems have begun to create formidable problems in that the volume of wastes has reached a point where even a light rain makes it impossible to treat all the sewage discharged by large communities. As an example, the city of Buffalo, New York recently experienced a storm lasting a mere fifteen minutes and producing less than a tenth of an inch of water, but the city's combined sewer system and water treatment facilities were taxed far beyond capacity. In addition, during the four hour overflow, the city had to discharge, untreated, nine times its normal load of treated sewage.

In former years the problem of an inadequate water supply for human use was not serious. This has all changed. Not only have metropolitan areas expanded beyond their water distribution facilities, but per capita consumption has grown several fold due to increased industrial and domestic use.[4] It is this increased demand (shift in the schedule) for water that has produced serious problems in many communities. Shortages, due to fixed (maximum) prices (see Appendix A) that currently exist, will become more serious during the next two decades and will more than likely affect communities now amply supplied. Inevitably, industrial and personal consumption of water will increase.

To meet this growing need for water, there is only a limited supply of ground and surface water suitable for domestic, industrial, and agricultural purposes. Although desalinization of ocean water may prove economically feasible within the near future, great reliance must still be placed on better development and more efficient utilization of fresh water supplies.

Unfortunately, disposal of the increased quantity of human and industrial wastes has seriously jeopardized the quality of water supplies and is rendering some of them unsuitable for human use. Unregulated urbanization and suburbanization and unrestricted growth of chemical technology have discharged wastes into our rivers and lakes and ground water facilities, making them into vast open sewers. Not only has this resulted in esthetic deterioration and the destruction of aquatic life, but it has also induced health hazards, whose potential significance cannot be ignored.

The cholera germ, [5] a curved, rod-like bacterium (*vibrio comma*), takes only two or three days to incubate in the body of a victim. In serious cases, afflicted individuals suffer paroxysms of diarrhea and vomiting, followed by collapse, delirium, and finally, coma. Death may occur in less than a day from the first evidence of illness, producing a ghastly, extreme form of rigor mortis.

Cholera can at least be attacked by effective water purification. The conventional forms of water treatment, however, do not necessarily prevent certain other pathogenic agents from affecting the public. Conventional water purifying techniques are effective enough against bacteria but what of the most sinister pathogens of all, the viruses? Despite many precautions, hepatitis has begun to emerge with prominence on the American disease landscape. The resulting jaundice of the skin and eyeballs and the deep yellow urine color are merely surface symptoms of liver injury. Fortunately, infectious hepatitis is rarely fatal. Experiments clearly reveal that the hepatitis virus survives conventional methods of water treatment. Ordinary chlorination of water simply lengthens the time required for the virus to incubate in the human body but fails to inactivate it. The virus, in effect, is weakened, but unless chlorine levels are unusually high, it is not destroyed. The most spectacular epidemic of water-borne hepatitis occurred in New Delhi, India, in December 1955, when an estimated 35 thousand people in an urban population of less than 2 million contracted the disease. Another example occurred following a flood of the Jamuna River. The water receded and the river changed its course, resulting in a flow upstream of raw sewage that caused gross contamination of the city water supply as it entered the water pumping station. It was calculated that during the peak of contamination about 50 percent of the water entering the pumping station was made up of raw sewage. Other epidemics of infectious hepatitis have also occurred recently in Sweden and the United States.[6]

Today, an unprecedented variety of cancer-causing agents enter the water supply. This is due, in part, to the ever-changing spectrum of exotic chemical wastes produced by our modern industries and agriculture. At least three elements—arsenic, atomic beryllium and chromium—are known to be carcinogenic. All three probably appear in the drinking water of industrial communities. Of course, the most important carcinogens in water are radioactive elements and aromatic chemicals. Aromatic chemicals are large groups of hydrocarbons with molecular structures resembling rings, as distinguished from ordinary hydrocarbons, which form long chains. Although most aromatics are not carcinogens, they include some of the most potent cancer-causing substances known to man, such as benzopyrene and benzanthracene. The majority of these aromatic carcinogens are produced by distilling or burning organic materials, such as coal and oil, and tend to contaminate the waters not only in industrial areas, but also in densely populated urban centers. Fish with cancers of the mouth, for example, have been found near the oil refineries of Los Angeles Bay, apparently because they had been feeding in the mud of water contaminated with refinery effluents. The carcinogen benzopyrene has been found in barnacles and oysters living in waters contaminated with ship fuel oil. The most widely known aromatic hydrocarbons found in water supplies today are pesticides, such as DDT, Dieldrin and Chlordane.

Taken individually, the effects of these pollutants are still very much a matter of speculation. No one quite knows what each one is doing to the human body.

Now imagine the world of tomorrow, if water pollution maintains its present course. The Atlantic Regional Fluid Control Board issued a favorable report after its yearly meeting. The meeting was held on the American continent for the first time in many years. Originally, the meetings were held alternately in Europe and America, but with the gradual buildup of solid banks of material in the Atlantic Fluid Area and the increasing viscosity of the remaining liquid parts due to the steady growth in heavy molecule effluents, wave action in the Fluid Area had damped down to very low levels. The Board, therefore, thought it a practical sign of regional solidarity to hold the meetings on a giant floating platform in the vicinity of the Mid-Atlantic Silt Ridge. During the last year or so, however, concentrations of noxious effluents have tended to accumulate near the site of the Board meeting, provoking complaints from many members. The engineering and maintenance staff of the Board also remarked on the acidic corrosion of structural materials in the platform caused by contact with the fluid, making the preservation of the site an expensive and difficult task. So the Board moved, reluctantly, to transfer the meeting place to dry land.

The good news in the Board report was that the Fluid Area, during the last five years, analyzed out at water content of 72 percent, compared to the previous five years' analysis of 69 percent. The increase was attributed to the imposition of newer, more stringent controls on sewage, agricultural waste, and, especially, dangerous industrial liquid wastes. Some experts also thought that the decrease in growth of marine microflora had led to an increase in the carbon dioxide in the atmosphere. This increase would then have led to a greater heating of the atmosphere and, therefore, more evaporation and precipitation and, consequently, better fluid quality. Other experts said that rainfall pattern changes had been too indefinite to draw such conclusions.

The decrease in algae and other marine flora was mentioned in another context in the report. Originally, the increased growth of flora caused by the injection of nutrients from fertilizers and other effluent components had seemed disastrous; the fishing industry had disappeared entirely from the economy. But biological scientists soon pointed out that the flora might be a valuable food source in itself. Experimental stations and pilot plants were set up, and, finally, several promising industries were established to convert algae and other microlife to foodstuffs. The Bread Upon the Waters Corporation proved an immense success with its stock, issued initially at $8 on the New York Stock Exchange, shooting up to $150 within three years.

Unfortunately, within the past few years this remarkable growth of marine plants has slowed, and in some cases in the Pacific and Indian Fluid Areas vast regions of algae have disappeared entirely. The exact cause has not yet been determined. Scientists believe that a number of chemicals present in the effluents from agricultural and industrial processes may be responsible. Pesticides, particularly, were suspected of having developed enough new unusually poisonous characteristics (when mixed together in large enough concentrations) to be fatal to the new marine industry

The Board report closed on a note of optimism, however. Recent experiments had shown that some important chemicals were occurring in such large quantities in the five major fluid areas of the world that reclaiming them in quantity was now approaching economic feasibility.

The world just described is, again, fiction. The tendencies are there, however. Who would have thought thirty years ago that Lake Erie would become a biological desert, destitute of many desirable aquatic species. We can no longer afford to be taken by surprise by water pollution disasters.

Poisons for All

The technological advances of this century and the next have been and will be introducing a new type of environmental hazard; i.e., artificial hazards of a chemical and physical na-

205

ture, all of which are the enviable consequences of modern scientific progress. Chemical and radiological technology will also produce new and ill-defined hazards. So we are presently undergoing a continuous process of artificial contamination of the environment. Many of these artificial substances are poisonous and their concentrations are increasing. Each new technological advance threatens us potentially with further contamination of the environment.[7]

Food is a potential vector of disease, as has been recognized for many centuries, and man has, therefore, adopted certain standard sanitary measures. Episodes of food-borne illness have dramatized the extent to which food is dangerous for human consumption. Modern transportation enables unsafe foodstuffs to endanger life at great distances from a single point of distribution. Highly toxic chemicals in agriculture, whether fertilizers or pesticides (insecticides, herbicides, or fungicides), may have harmful effects on those who consume the foods. Their toxic effects on agricultural workers or airplane pilots, who are poisoned while applying these chemicals, are well documented. We can also wonder about the cumulative effects of current or increased levels of ingestion. The consequences of new chemicals that inevitably will be used in future years must also give us pause. Farming methods are, of course, modifiable, but there is no present reason to anticipate that agriculture will soon revert to the methods of our forefathers. The needs of high yield require the continuing use of current fertilizing practices and chemical control of pests and plant disease, and, unless these threats are controlled, the communities of tomorrow will be faced with a hazard magnified many times from present-day agricultural practices.

The supply of meat and animal products also entails both recognized and potential health hazards. Trichinosis, tapeworm, and salmonella infections from the use of meat from infected animals and the hazards incidental to improper or careless practices with respect to milk and dairy products are well recognized risks. Various antibiotics have been used to treat infections in cattle, and the result has sometimes been the contamination of milk with these substances. Hormones used to accelerate animal growth are transmitted through the consumption of meat, and demonstrable concentrations of certain pesticides used to control insects around stables have been found in milk. We still cannot fully evaluate the significance of these practices, but their potential import cannot be ignored. Complicated biological and technological processes can be controlled only through equally complicated technical tests and procedures supplied by trained personnel. Such personnel are just not available to small industry or to small units of government.[8]

The biological hazards of environment include all forms of life which, through association with man, may act to the detriment of the biosphere. These include not only pathogenic microorganisms, but also the animal reservoirs from which they may spread and the insects or other animals that participate in their migration. They include plant life producing pollens to which man is allergic and wild animal life and toxic plants with which man may be in contact in recreational areas. These latter hazards are potentially serious to many individuals, but at least they have the virtue of familiarity to us sufferers from hay fever and poison ivy.

Insects sometimes carry disease. That is bad. Unfortunately, extensive use of insecticides to control, for example, mosquitoes as nuisances has led to the development of insect resistance, which increases the difficulty of control when it becomes critically necessary for the prevention of the spread of disease. This is also bad—maybe worse. Over a hundred different animal infections have been known to attack man. While many of these do not, at present, exist in the United States, one can definitely count on their introduction in the not so distant future. Some of these may infect wildlife and may spread to man directly or via domestic animals or insects. Others are basically diseases of domestic animals. A tremendous increase in the sale of household pets, including cats, dogs, birds, and turtles, has brought with it a close human association with animals, many of which carry infection into the home. Rabies from dogs, psittacosis from parakeets, and salmonella infections from pet turtles are common examples of such conditions. Even more significant are the infections that come via food from infected animals, notably the trichinosis and salmonellosis mentioned above.[9]

In the past twenty years several million pounds of mercury have been dumped into the national waterways.[10] It was assumed that mercury would sink to the bottom of lakes or rivers, but such is not the case. Mercury, officials now realize, changes its potency in water from a harmless inorganic form into methyl mercury, a very deadly form. Microorganisms, which are eaten by small fish, which in turn are eaten by larger fish, pick up minute amounts of methyl mercury. If humans eat the contaminated pike and pickerel over long periods of time, they can receive dosages large enough to cause blindness, brain damage, and even death. And then there are the new found environmental hazards of lead, cadmium, and nickel carbonyl.

When we try to imagine a world with too many poisons, we may do well to visualize the future as a kind of retrogression to the past. To a Westerner, the appropriate retrogression is to the Europe of the Early Modern Age, when black plague, tuberculosis (white plague), and cholera were an unwelcome but accepted part of life. To the Third World, the return to the past is to the very recent past or, in some cases, to the present.

The prospect is discouraging since medicine has progressed so far. From an understanding of the bacterial diseases, there is now developing a growing awareness of the role of resistance to infection. There has even been progress in understanding and curing diseases resulting from improved sanitation, such as poliomyelitis and, perhaps, multiple sclerosis.

The world of microbes does fight back. Physicians must now deal with antibiotic-resistant strains of pneumonia. There we at least have some idea of where the enemy will come from. On the other hand, when we add misunderstood chemicals to our ambience, we put ourselves in the position of not knowing what symptoms to expect, in addition to the usual problems of cause and cure.

This kind of sick future may bring on a new, almost medieval mentality. In this kind of world death and sickness will strike pandemically. Hopefully, many of the artificial diseases will be identified and the offending chemicals eliminated quickly, as has happened often during the past few years. But what about the long-term effects of such compounds as carbon tetrachloride, which has been discovered to cause liver damage? Whole populations may discover that they are suddenly condemned to invalidism or premature death. Once such epidemics have become common, a medieval sense of resignation may enter into the social psychological environment. The consequent feeling that life is cheap might have disastrous consequences. It seems possible that the formal institution of war is acceptable in society largely only to the extent that people feel the pressure of eventual mortality. An increase in suddenly discovered mortal illnesses could well contribute to a devil-may-care attitude toward life, and consequently a greater acceptance of such institutions as mutual destruction. Now when weapons are excessively destructive, any change in making war more acceptable as an institution is tragedy (or sick comedy, depending on one's tastes).

All that can be said is, caution! Regulations must be severe on the introduction of new additives. The public good is subject to great danger from this problem of "poisons for everyone." We cannot afford leaving the decisions, even in part, to private agents. Any reasonable cost increase necessary for the beefing-up of regulatory agencies must be accepted by the public, if the poison disaster is to be averted in time.

Too Many People

One-third of mankind lives in the city today, and this proportion is increasing at such a rate that, *ceteris paribus*, by the middle of the next century some 95 percent of mankind will be living in the city.[11]

The effect of large urban centers on mental health and human behavior has been well established.[12] Roughly one out of four people in Manhattan are reported to have neuroses sufficiently severe to disrupt their daily lives. These figures may be compared with the currently accepted figures of the country as a whole, where roughly one out of ten Americans in all age groups suffers from mental illness severe enough to warrant treatment. Observations of animal colonies allowed to multiply in confined areas have shown for each species a critical limit to the number of animals that can normally function in a given space. Once this limit is exceeded, the social organization of the population breaks down and destructive and antisocial behavior predominates. Certainly, man also has critical space requirements and his reactions to the psychologic and physiologic pressures and frustrations of crowding and inadequate living space can result in social withdrawal or civil disruption.

The cubical life of a human honeycomb is something akin to living in an African drum. Every household beat reverberates so that no one literally knows a moment of silence or true privacy. In a normally quiet suburban residential area, sound levels range between twenty and thirty decibels. They are more than doubled in a Manhattan hotel room—from fifty to sixty decibels. Aside from the psychological effects from noise, which may be defined as unwanted sounds, there is no longer any doubt that it impairs the hearing of individuals who live or work in noisy surroundings. Of course, excessive noise may have even more far reaching effects. Scientists find that the sudden slam of a door may raise the blood pressure appreciably above its normal level. Laboratory rats subjected to sirens have developed gastric ulcers and suffer from severe overstimulation of the adrenal glands.

Perhaps the most fantastic combination of noise and congestion is reached in the great transportation arteries of sprawling cities in the Western World. The ear-splitting effects of this experience reach appalling proportions; that is, traffic-noise levels reach in excess of one hundred decibels. We speculate that there are even higher levels in Los Angeles because of the buying mania for the motorcycle, another new toy. Added to this is the fact that during certain hours of the day everyone seems to be going to the same place at the same time. Vehicular progress in Los Angeles has dropped from a galloping rate of approximately twelve miles per hour in the horse and buggy era to a crawling six miles per hour during peak driving periods (assuming no accidents) in the jet-propelled era. All such overcrowding is a distinct insult to the nervous system.

Overcrowding creates its own kind of atmosphere, in the literal sense. It is a well known fact that the motor car, in graduating from a curiosity to a routine means of transportation, has become a major source of air pollution, filling the atmosphere with an appalling array of gaseous and solid poisons. The innocent looking tailpipe of an automobile spews forth about 200 hydrocarbons, some of which are provisionally identified as cancer causing substances. In the course of consuming 1,000 gallons of gasoline, motor vehicles typically discharge 17 pounds of sulfur dioxide, 18 pounds of aldehydes, 20 to 75 pounds of oxides of nitrogen, and more than 3,200 pounds of carbon monoxide.[13] Sulfur dioxide is not only a well-known irritant, but a poison. Many aldehydes,

211

particularly formaldehyde, are corrosive agents. Certain oxides of nitrogen are toxic even in very small quantities, and carbon monoxide, perhaps the most familiar of this group, competes favorably with barbiturates as a means of achieving a lasting self-delivery from the burdens of the world. Many an Angeleno, after submitting to sharply curtailing domestic and industrial sources of contamination, is disillusioned to learn that as much as 80 percent of the pollutants are produced by the city's privately owned motor vehicles. We are not aware of any subsequent curtailment in automobile use. But we have, of course, already considered this "overcrowded air" as a special problem of its own.

All this and more can be considered as a function of too many people. A continued gap between the birth and death rate results in a geometric increase in the total population. Migration into already densely populated areas can only add to all the problems. And given the present state of technology (or even changing it a little), continued population growth can result in pathological determinants of future population growth and characteristics. It would seem reasonable to argue that quality, not quantity is the sine qua non for any economy that seeks to conserve its environment. Unless something like the Malthusian negative checks can be promoted as a means to an end, coupled with the modification of present-day consumption habits and patterns, an adaptation to the polluted environment may be enforced as a rule of nature.

We have spoken in Chapter 10 about the possibility of a viable community on an overcrowded planet. If pollution is controlled and public needs are properly represented in the marketplace, there is no reason why life cannot be sweet (though gregarious). There is also no doubt that life will be different from anything we are used to.

Imagine a world of 50 billion people; living space is organized in much the same way as apartments in large cities, only more so. From the 3,500 foot high Tower of Tolerance, one can look at the block after block of concrete and steel dwelling units. Down below, the ground is covered with concrete pathways. These pathways are not, of course, roads for motor vehicles, internal combustion engines having long been outlawed. In fact, all travel is greatly restricted. Since space is at a premium and movement is hampered by congestion, the costs of transportation are extremely high. But significant economies of scale for localized enterprises can be easily realized because large markets for products are available in the small but very densely settled geographical living units. So most people are able to satisfy their consumption needs within their local neighborhood. This natural tendency for the decrease of long-range transportation has been reinforced further by public policy actions and regulation. Public air, space, and land transportation facilities have been steadily decreased by the withdrawal of franchises, and fare rates for interneighborhood transport have been set at such high levels that only the very rich or the very powerful can afford to travel. Travel by private electric mini-vehicles is also discouraged. Road tolls are high and the former networks of freeways have been gradually broken up by removing key links between neighborhoods. These discontinued links were formerly sold to private developers for building purposes. There has been public pressure recently to devote more land to park use, and a recently passed law provides that every neighborhood of over 250,000 inhabitants is to have space for five large half-acre parks set aside from freeway discontinuance land condemnations.

Nature enthusiasts have applauded this reversal of the long-time trend toward less and less open land. The Save the Sunflowers Society (a lineal descendant of the Save the Redwoods Society) has also preserved, against protests from inhabitants of crowded ghettos, a striking stand of five acres of sunflowers, dandelions, and petunias, which is open to the public in the Kansas region. Whether or not they will be able

213

to save this unique botanical ecosystem from the pressures of the League of the Underhoused, with its influential lobby in Congress, is another story. The League spokesman correctly pointed out that the five acres of land could provide better housing for 10,000 people, who are now jammed into inadequate residential units in the older sections of the Garden City, Kansas Neighborhood No. 204. The point was made that trips to the Preserve are only practical for the rich, who often have their own window boxes (usually shoeboxes of geraniums) at home.

The nature lovers have a strong supporter in the President, who is credited with having used his personal influence to save a sample one-mile stretch of the Brazos River in Texas as a National Monument. Water resource specialists objected that making an exception in the riverbed reuse program endangered the Texas section of the nationwide water catchment and underground conduit system, in addition to wasting valuable housing space.

The President, on the advice of his Department of Unrealized Housing, also recently instituted a nationwide lottery scheme tied to the purchase of government savings bonds. Winners of the lottery are given free tours of the National Monuments, including not only the Brazos River Sample and the Sunflower Preserve, but also the Giant Redwood and the Rocky Mountain Peak Summit. In this way, it is hoped that ordinary members of the public can identify better with the Monument System.

The bond lottery system promised to be a great success in other ways. With the gradual shortening of the work day during the past decades to one hour and forty-five minutes, gambling of all kinds has been popular, and the lottery is a welcome addition to cards, dice, and guinea-pig racing as an attraction. In addition to the tours and the usual cash prizes, the government has also been able to provide unusual prizes not ordinarily available to private or semipublic concerns.

Only the government could guarantee an all-expense-paid vacation for one, in which the prize winner is entitled to spend as much time as he wished for a period of two weeks, absolutely alone in a vacated federal warehouse.

It may come to this. It may come to worse, with wars and famines taking up again their classical roles as population controls. Whether this world of the future is viable or not in the first place depends on rationality in planning and decisions, not just on population size alone. Whether a viable world of enormous population is sick or not, will depend on personal taste. It is strange to us, but would not the world of today appear bizarre to King Nebuchadnezzar of Babylon?

Getting Back on the Track

Continued increases in the hazard of pollution associated with the water supply contamination, liquid waste disposal, air pollution, radiological hazards, solid wastes, food additives, animal and plant hazards, pesticides and noise, and the psychological dangers to the well-being of man, seem sufficiently obvious that man should be moved to do something about this problem. In other words, we should get back on the track. This is obviously easier said than done. The social, economic, and technological changes that have made possible the removal of the most important environmental hazards of former years have at the same time introduced new and potentially more serious hazards. We now realize that the water we drink, the food we eat, and the air we breathe contain increasing quantities of chemical contaminants of a potentially harmful nature. Nuclear power, which offers one hope as a source of energy for the future, has introduced environmental hazards that were unknown and not envisioned by former generations. Environmental hazards that were at one time purely local are today problems of interregional and international concern and, therefore, must be solved on a different political basis than the problems of former years.

In order to get back on the track, it is important to identify both present and probable future environmental hazards and to measure their relative importance. We must actively pursue optimal ecostabilization policies with both the resources that are currently available and those that will be required for future environmental programs, and we must assess in money terms the effects of proposed policies by making use of modern economic tools and techniques for evaluating and recommending ecostabilization programs.

12

A PROGRAM FOR LIFE:
ECONECOL

The ecological crisis calls for some kind of programmatic approach. It may be, of course, that we can muddle through this crisis as we have through past difficulties. Certainly no kind of program is perfect anyway, so it is important that our democratic (and centralist, in the socialist countries) planners be skilled in improvisation. But how much better it is to have in mind some kind of methodology or way of proceeding. We call this methodology, or program ECONECOL, borrowing the syntax methods of the computer programmers as a shorthand labeling device. Some idea of the nature and extent of this program must be evident from what we have said so far. We would like to summarize some concrete suggestions here.

The emphasis, of course, is on government planning. This source of aid for the environment is distasteful to some, but we do not see, in this post-Keynes era, how a more plausible choice can be made.

Perhaps popular mistrust of bureaucracy does contain some elements of unrealistic nostalgia for the good old days, or for imagined Golden Ages of the past. The bureau, after all, appears to be a necessary institution in the complex world of today. One recent study [1] lists several reasons why bureaucracy is here to stay. Most importantly, the provision of "indivisible benefits," or "collective goods" is one of the most prominent functions of bureaus, as the size of the Defense Department budget attests, and even pacifists will recognize the reason for the existence of an educational bureaucracy at least in the present centralized system.[2] The Internal Revenue Service is an example of another function, the use of nonmarket hierarchical organizations to perform redistribution of incomes. Other recognizable important functions of bureaus are such desirable actions as the regulation of monopolies, the protection of the consumer, the stabilization of markets, the maintenance of law and order, and the internal housekeeping of the government itself. Also, we must not forget the role of bureaucracy in trying to manage large externalities of benefits and costs, since the regulation of pollutant processes is very much our concern here.

Certainly, if all social benefits were marketable, bureaucratic organizations could be done away with.[3] An example of this kind of process might be the recent shift of the post office from a government department to a public corporation. Even in this example, there arise questions about subsidized classes of mail and the Rural Free Delivery service. Do such money-losers actually represent indivisible benefits to us all, or do they not? If so, a special bureaucracy to supervise the handling of such indivisible benefits would seem an unescapable feature of postal operations. Similar problems arise in schemes to set up a voucher system of education. It would seem likely that a completely free educational market might successfully offer inferior education to some children, and the general attitude toward education appears to be that the bad education of some harms the entire society. So some bureaucratic regulation would always be necessary to insure proper quality standards.

When it comes down to the regulation of monopolies and

compensating for the deficiencies of the free market, there appears to be no marketable component in present bureaucratic functions. And so far as we can see, there is little or no directly marketable component in diseconomies such as smog. Try as we may, we cannot, at present, buy clean air. So if a bureaucracy is needed to modify the market to recover the costs imposed by smog, we view it as, at most, a regrettable necessity.

Others have taken this same point of view. Eminent sociologists have viewed bureaucracy as an "evolutionary universal," and as the "most effective large-scale administrative organization that man has invented," remarking that "there is no substitute for it." [4] If in the future substitutes for bureaus can be found, well and good. In the meantime, let us work with what we have.

One Universe

The scope of the ecological problem is world-wide. Smog is common to New York, Paris, and Tokyo. Water pollution experts in the United States hope to learn from the experience and regulatory schemes of the planning authorities in the Ruhr Basin in Western Germany. Erosion and wastage of land resources are as old and universal as agricultural man himself. Most of these events are common and localized experiences, rather than true international interactive processes. They do form a common bond of sympathy between the New Yorker and the Muscovite, but they do not require actual international cooperation. It does not take a great deal of imagination, however, to visualize times in the near future when air pollution will become a truly international problem. Water pollution, of course, is already recognized as one, as in the case of the fish poisoning in the Rhine River which occurred not so long ago.

We might then suspect that international cooperation in the ecological problem will become necessary soon. Of course, cooperation, in the sense of setting up an international commission and issuing reports detailing the destructive things that are happening to the environment, should be easy to come by, but any real cooperation to establish a methodical program which attacks the environmental disturbance problem seems to have rather discouraging prospects. After all, we have only tried to emphasize the difficulties in securing rational planning at a national level. Difficulties at the international level would be that much more severe.

Several particular problems spring to mind. At the level of national planning, there is a commonly accepted indicator of overall social output or level of economic success, the gross national product (GNP). That product must be modified by including the externalities we have talked so much about, if it is to suit our planning needs. The corresponding indicator on the world level would be a modified gross world product (GWP). This kind of formulation presupposes that all nations will be considered equal. History, unfortunately, shows that one way or the other some nations end up being more equal than others (a George Orwell phrase). So there may be quarrels between nations on how each country's productive output should be judged in relation to others as they enter into planning goals, and when there is no agreement on goals, there can hardly be any agreement on means. But agreement on at least partial goals could lead to some action on tentative policies, and agreement on some kind of policy is, oddly enough, even more important than agreement on goals. Even a poor policy would at least teach us what not to do. No policy means no change and a drift toward self-destruction through apathy.

Hopefully, such difficulties can be solved by compromise. By analogy with the Common Market negotiations on trade agreements in Europe, it might be possible to trade off optimization goals in different economic sectors for different countries. Agriculture might receive more weight for Cuba, and industry might receive more weight for the United States, to mention two countries with somewhat different viewpoints on world politics and economics. Certainly, machinery now exists with the United Nations which could be used to start working on such an international program, if the will and the money are there.

The reference to Cuba reminds us of a difficulty which is found at a different level of thinking. We have talked about the importance of manipulating the free market mechanisms to include the effects of environmental disturbance. We have pointed out that even in the Soviet Union the mechanism of the free market is being touted as a method of carrying out price and cost accounting for a socialist system. But this is far from asserting that the same methods of government action recommended for the United States will work for the Soviet Union. When one considers the problem of China, the contrast is even more marked. Of course, different methods of control can be used in different nations, all under one international scheme. But the presence of ideological restrictions in economic thinking is certain to be a trouble spot in any system of international cooperation. East is East and West is still West, even though economics and politics (communism vs. capitalism) may have partially replaced race and culture as points of division.

All of these problems might be included under the heading of trying to find out how to apply rational procedures to the planning of human destiny. Man is an emotional animal, but his problem in every sphere of life is how to express those emotions, while, at the same time, not letting them drive him to destroy his life patterns and the things that he wants and loves. The economist Milton Friedman points out that disagreements among people about means to a goal are often displaced, through intellectual laziness or confusion of thinking, and appear as disagreements about goals. Naturally, we should certainly be ready to admit that people differ, and that we have to coexist in the world with people having different goals of life in mind. But this truism is often more honored in the breach than in the observance in international relations. It has been remarked that if a person acted as a typical nation does, letting its actions be influenced by irrational national pride, fear of losing face, and readiness to go to war because of insults to national honor, we would describe that person as hopelessly neurotic. The phenomenon helps to complicate all kinds of negotiations on the international scene and would also complicate the handling of ecological problems in the present world political system. Indeed, this "irrationality syndrome," which makes paranoid behavior with regard to one's fellows an acceptable way of life, is, unfortunately, even seen in the domestic arena, as far as environmental problems are concerned. Such titles as *America The Raped* are examples of the kind of emotive terminology often used in discussing the ecological problem. Of course, playing on the emotions of the reader is an accepted facet of polemic style. Certainly, there is no terrible harm in name calling and trying to enlist public emotion on the side of just causes, but we fear that such strategy often backfires. All too often, the first round of the fight goes to the contestant who shouts louder; but the last round is won by those who really have the most at stake, independent of their skill in public debate. Such has often been the case with consumer groups vis-à-vis industrial firms in the United States. Consumer groups often appear to make progress in such fields as product standardization and truth in advertising, but the industries, who have a true "gut" concern with making profits from production, end up coming out on top. In some ways, this has been the history of the air pollution problem. Automobile manufacturers and oil firms have given ground, but the total result so far seems to indicate what one might suspect in the first place; the manufacturers are those most closely concerned, and they are the ones who tend to end up determining the main course of events. This fact has important implications for the problems of politics and power in the ecological context.

Politics and Power

Power in a democratic society is wielded partly by republican political institutions, and partly by pressure groups within the society. In planning for better ecosystems, we look to the political institutions for relief and guidance. We hope they will pass laws forbidding dangerous pollution and, in general, the degradation of our environment. We also suggest that government use its spending powers to establish such ecostabilization efforts as water quality schemes and research on solid waste management. We hope that sensible plans to levy compensatory taxes on diseconomy-producing industries will be carried forward. But we would be extremely naïve if we hoped to do this without taking into account the other nongovernmental centers of power. For our purposes here, one of the most important centers is the organized lobbying activities of large industries. As we have said, industry often wins pollution battles, not only because of its great financial resources and consequent political influence, but, in a more basic way, because the pollution question is a dollars-and-cents problem to them so that they have an extraordinary will to win. To cast industry in the role of villain or devil in this situation turns out to be a self-defeating tactic. Often producers win environmental struggles because their opponents are weighted down by extra ideological baggage that does not belong with the point at issue. Too often, the thinking of the environmentalists contains certain hidden assumptions, such as "capitalism is bad, anyway," "let's really make things tough

223

for the industry," "the producers are trying to squeeze every last penny of profit they can, let's pay them back good," and so on.

We believe that such ideological baggage should be thrown out of the environmental arena. Producers are people too; their motivations are quite understandable, and the desire for profits is both legal and moral under our system of government and economics. Unbridled pollution is immoral and should not be legal under any system. It is in the interest of the industrialists in the long-run to help prevent ecosystem degeneration, if only because they are citizens of the world too.

We need to speak softly as men of reason and carry the big stick of sensible manipulation of the market. Less name calling and more sensible government actions are needed. Instead of showy regulations that are not enforced, more modest laws with real teeth in them are needed. Enforcement efforts to detect large emitters of pollutants are often frustrated by the timing of emissions to coincide with the absence of inspectors. Laws setting small fines for relatively small emissions could establish a pattern of compliance, since evasion would be less worthwhile and perhaps more unpractical. If regulations are not appropriate, scrupulously fair taxes should be assessed on producers of pollutants.

A first practical step in getting a down-to-earth program started would be the setting up of a joint committee of Congress to work with an appropriate executive agency. Such a legislative joint committee and executive agency would be charged with setting up plans for the "piecemeal attack" on ecosystem problems. It would have the task of educating the public and the affected industries in the methods and goals of the plans.

The atomic energy problem is a very good example of one effective way to get things done. With the atomic bomb, of course, one had the advantage that the weapon aroused tremendous fears, so drastic powers were thought appropriate and were indeed given to the Joint Congressional Committee on Atomic Energy and the Atomic Energy Commission itself. Perhaps we might have to take back a few of our words about rationality here and say that it might be a good thing if people were, after all, a little more afraid of the environmental problem—afraid in a constructive way, of course.

Naturally, the expertise already represented by the Federal agencies and having to do with the environment would be invaluable. Currently proposed and recently implemented reorganization schemes for these agencies might form an essential step in attaining the goals outlined here. But if these schemes give us just the same old bureaucratic reshuffling, with insufficient power and intelligence to use it wisely at key decision points, they will be worse than useless. The "Superdepartment of Environment" proposed here requires much more than that.

Of course, wishes are not horses, and it is easy to require "power and intelligence to use it wisely," but much harder to carry it out in practice. Perhaps the system is incapable of such steps. We do not think so. We do not think that the disagreements about methods and the conflicts of public and private interest which often impede needed reforms are to be equated with a general incapacity. If the system is incapable of specific reforms, that itself may perhaps be a sign of general societal degeneration. And for degenerate societies, the barbarians usually await both within and without the gates. Their presence there today does not need to be emphasized.

Can these disaffected external and internal proletarians [5] of our civilization hope to do better themselves? We can observe with interest some of their experiments in new ways of living, but we cannot speak for them. We can only note that often they themselves appear to vacillate between two contradictory hypotheses—change inside and outside the advanced industrial society.[6] So let us stick to our conventionally-

225

phrased recommendations for action within the present political system and consider the role of a "Super-Department of Environment."

Educational and Information Flow

The proposed ecological agency, or "Super-Department of Environment," would have three main functions connected with research and information flow: (1) A public information department would have responsibility for presenting the facts on environmental disturbances in the country to the public in a fair and unbiased manner. The methodology behind the planning efforts of the commission should be explained in simple terms to the public. Both the scientific data and the economic facts of life deserve the best efforts of experts in mass-media communication in publicly presenting the situation. (2) A scientific research department should be set up to overlook and direct contract support of scientific research slanted toward the solution of ecological problems. (3) An economic research department should carry out in-house analyses of the economics of pollution and other forms of environmental disturbance. It should also administer contract support to the universities and firms specializing in the economics of the environment.

Planning, Assessment, and Reassessment

The key to the problem is finding the people to do the planning job, and getting them the support they need to carry out the work. Presumably, if the "Super-Department of Environment" does its job in educating the public to the needs of the moment, they will get enough money to enable them to guide all of us in saving ourselves and our environment. Hard decisions will have to be made, but in line with some such methodology as is discussed in Chapter 9, decisions can be defended on rational, if not infallible reasoning. We can guess that a lot of toes will be stepped on if any effective regulation plans are created, just because the problem intuitively appears to be so enormous. So it is essential that the department be careful in optimizing its results, that is, in testing recommendations against possible errors caused by our ignorance of what exactly is going on in the economic, not to mention the technical sphere. Pure politics will also play a part in agency decisions in the real world, but presumably the joint committee of Congress is supposed to help cover that part of the infield.

The crux of the matter is the use of a method. The method, to be useful, must deal with agreed upon values. Those who use the method must not take advantage of it. Man is a part of the ecosystem; he shares many imperfections with it, and his brainwork is liable to errors, both of fact and interpretation. As in all human undertakings, the fight against environmental disruption calls for a decisiveness of approach, but it also demands carefulness and flexibility in execution. We realize our modest place in the universe and our rather immodest ability to foul our own nest, and the ecological crisis has already reinforced our feelings of humility. It would be a shame to lose that all-important grace in the fight to protect nature and ourselves from the evil consequences of our own genius.

APPENDIX A
A NOTE ON DEMAND
AND SUPPLY

We introduce this note to sum up the behavior of firms and individuals in terms of the market mechanism: supply and demand. Well-meaning but economically unsophisticated planners attempt to bring economics up to the high state of their discipline by bandying these two words around. The unemployed, as well as the employed read about the demand for labor, written by an individual who may or may not know the difference between a demand curve and a point on the curve or a shift in the schedule. We hear about things being in short supply. What does that mean? Out there are millions of learned political economists—parrots with their vestige of feathers from a previous age. To paraphrase a well-known quotation: "Everybody talks about supply and demand, but only a few of utility and costs." We present this material with no diagrams or formal mathematics.

Demand

Let us start with a review of market demand as developed by Alfred Marshall (1842–1924), an economist's economist whose theoretical brilliance still radiates from his book, *Principles of Economics*.[1]

Demand is the total of the demands of the individual buyers in a market. It may or may not be a competitive market. For simplicity, we will assume it is competitive, i.e., that the product sold is of the same quality (homogeneity); that the product may be bought in either large or small quantities (divisibility); that there are a large number of buyers and sellers and neither can appreciably influence the price of the product (pure competition); and that there are no transportation costs (perfect market).

In a given market in a given period of time, the demand function for a product is the relation between the various amounts of the product that may be bought and the determinants of these amounts, including the prices of the product, the individuals' incomes, the prices of other products (substitutes and complements), and the tastes or desires of the individuals. By employing the *ceteris paribus* clause, the three other determinants of demand—incomes, prices of other products, and tastes—are held constant in order to focus on the fundamental relationship between demand and price, a relationship that takes the center stage in economic analysis.

If one plots the points of a *demand schedule*, which is nothing more than a list of prices and the quantities demanded at those prices, then the price-quantity relation is illustrated geometrically and is called a *demand curve*. We have assumed that price and quantity vary continuously, and with unimportant exceptions the demand curve will have a negative slope, i.e., there is an inverse relationship between price and quantity. This is called the *law of demand*, the firm logic of which is supported by the law of diminishing marginal utility (neo-classical cardinal analysis) and by indifference curve analysis (the ordinal approach which is not developed in this book).

To repeat, any point on the demand curve states the quantity that will be demanded at the given price. So when we speak of the *quantity demanded* we are speaking of a particular point on the demand curve. When we speak of demand we are talking of the entire demand schedule for that product. It is important to understand this communication distinction. A change in demand (not quantity demanded) is a shift in the demand curve—a change in the entire demand schedule. If demand decreases, all of the quantities opposite each of the prices become smaller. When demand increases, consumers are willing to pay a higher price than before for any given quantity. The causes of shifts in demand may be attributed to a rise or fall in incomes, prices of other products, and weaker or stronger desires.

Elasticity of Demand

The elasticity of demand (sometimes called price elasticity) is a technical term, formulated by Alfred Marshall to describe the degree of responsiveness of a good to a fall (or rise) in its price. The standard definition is as follows: *Elasticity of demand equals (minus) the relative change in amount demanded, divided by the relative change in price.* The elasticity concept is useful as a shorthand for predictions of whether or not changes in price will have much effect on changes in demand. The usual words used in this connection are *elastic* or *inelastic*. By definition, if the elasticity is greater than one, demand is *elastic*, i.e., a given fall in price will result in a relatively larger increase in the quantity demanded. If the elasticity is less than one, demand is *inelastic*, i.e., a given fall in price will result in a relatively smaller increase in the quantity demanded. If the elasticity is zero, demand is *perfectly inelastic,* or a vertical demand curve. If the elasticity is infinite, demand is *perfectly elastic,* or a horizontal demand curve. All elasticities, when measured numerically, will lie between these two limits. If the elasticity is exactly equal to one, demand is said to have *unit elasticity,* i.e., a given fall in price causes no change in total expenditures. A rectangular hyperbola (price inversely proportional to the quantities demanded) exhibits this property.

Economists often work with a particular type of demand curve that makes calculations and conceptualization relatively simple. This *linear* demand curve supposes that the quantities demanded goes up in proportion as price goes down. Since the absolute changes in price and amounts demanded are proportional, the elasticity is only a constant divided by the quantity demanded and multiplied by the price. For this linear case, which is often used in illustrations, elasticity is small for low prices and large amounts demanded, but when prices become sufficiently high and amounts demanded sufficiently small, the elasticity becomes large.

This change in elasticity for linear demand curves is significant for economists, business firms and government planners. The total expenditures (or receipts) involved will go up if the price is raised when the demand curve is inelastic, but will go down if the demand curve is in the elastic region. This comes about because the total expenditures are just the price times the quantity demanded and inelasticity means that the relative demand indeed falls, but not proportionately to the relative price change. So unit sales fall off as prices are raised in the inelastic region, but the price rise more than makes up for it, and total receipts rise. When, finally, the demand becomes elastic (in the linear case, for high enough prices), relative demand falls off faster than relative price, and the seller loses out on receipts by raising prices. And if unit costs are constant, smaller receipts means less profit. So it is good for the businessman or for the government body taxing him (perhaps for ecostabilization purposes) to be aware of changes in elasticity as prices rise.

The Congress of the United States is well aware of the concept of elasticity and, in fixing taxes, tries to determine which commodities it should seek to tax to get more revenue. Inelasticity of demand also explains what economists call "the paradox of plenty." For example, take a particular product in the agriculture sector for which there is inelastic demand. A bountiful crop results in a smaller total revenue to its growers. Or if the demand of a particular farm product were elastic, any effort by the government or the growers themselves to raise the price of the commodity would result in lower gross incomes.

Supply

The other cutting edge of the economics scissors is supply, and the general rule is that supply curves are positively sloped upwards from left to right. This means sellers of a good are willing to sell more at a higher price and less at a lower price. It is the world of the firm and its output decisions, where the firm (or producer) has the alternative of producing or not producing, or of covering costs or not covering costs.

The main factor determining supply prices of varying outputs is the cost of producing them. In order to produce some output, two or more factors of production must be employed e.g., land, labor, capital, and entrepreneurship. The principle is simple: if consumers want more they have to pay more for it. The reason is that the firm will have to attract more factors of production away from other industries. These factors may be more expensive and possibly less efficient, or both. A clas-

233

sic example may be yesterday's aerospace industry where (ignoring the "Peter Principle") firms within the industry com-
peted for labor factor inputs (engineers) by paying wages in
excess of the marginal contribution of the employees, raising,
appreciably, the cost per unit of output. Also, we run into the
"law of variable proportions," or *diminishing returns,* i.e.,
given the state of technology, as additional variable factor inputs (e.g., labor) are added to other factors which are fixed
(e.g., land) the additions of output in physical terms eventually decrease. This means that each additional input adds less
output than the previous unit. Obviously, if the price paid for
each factor input is the same (or increasing as in the aerospace example) and each additional factor level adds less to
total output then costs (marginal) must eventually increase.

This explains why supply curves slope upwards and to the
right, and it is safe to assume that the higher the price of any
good, the more will be supplied, as long as the firm is covering its alternative costs, including normal profits for the industry.

Again, you should think in terms of a *supply schedule,* or
curve, which shows the relation between market price and
the amounts of the good that the producers are willing to
supply. When we speak of *supply,* we are speaking of the
whole schedule or curve; whereas *quantity supplied* is a point
on that curve. An increase in supply is a shift of the curve
downwards and to the right, e.g., cheaper corn cost makes
farmers willing to offer more hogs at each of the different
prices shown. A decrease in supply is a shift of the curve upwards and to the left, e.g., a bad harvest results in farmers of-
fering less corn at each of the different prices shown.

Elasticity of Supply

The elasticity of supply is another technical term used by economists to measure responsiveness, i.e., whether one supply curve is more or less elastic than another supply curve. The general measure is: *Elasticity of supply equals the relative increase in amount supplied, divided by the relative increase in price.* Since both price and quantity usually move in the same direction, the elasticity coefficient has a positive sign.

Again, there are two limiting cases. The supply curve is perfectly inelastic when it is a vertical line, i.e., the quantity supplied is totally unresponsive to any price changes. Supply is perfectly elastic when it is a horizontal line, i.e., a small rise in price evokes an indefinitely large increase in the amount supplied. Unresponsive supply is defined as zero elasticity, and infinitely responsive supply as infinite elasticity. The degree of responsiveness of the amount supplied falls between these two limits. Unitary elasticity of supply would be represented by any supply curve through the origin, and whatever the scales of the two axes, the elasticity coefficient will always equal one. The reason for this is that the geometry of the elasticity of supply is different from that of the elasticity of demand. There is no important economic significance, in terms of total consumer expenditure, to be attached to the elasticity of supply, except to zero and infinite elasticities. However, the distinction between small or great supply elasticities is important, as will be seen below when reviewing the "incidence" of a tax.

Supply and Demand

The two forces of supply and demand, operating in our theoretically competitive marketplace, will establish prices (or

value, or worthiness)—prices that will induce production
and prices that will restrict consumption; prices (relative) that
will induce producers to select one production method over
another, and prices that will encourage consumption of one
thing and less of another; prices that indicate where research
is needed to reduce costs, and which of any number of possible new methods is best; prices established by the two forces
of supply and demand that guide, or allocate scarce resources
to society's advantage. It is the "law" of supply and demand
which is the important cybernetic principle that economists
have developed to explain the economic activities of people.

Equilibrium

Continue to assume that the demand and supply functions
are linear, and they are brought together in a competitive
market. What now follows is a partial equilibrium price, a
point where there is no net tendency to move, as long as
basic conditions are unchanged. Should there by any price
other than the equilibrium price, there will be forces for
change toward a price at which the quantity supplied is just
equal to the quantity demanded. That is, buyers and sellers
can benefit by doing something differently and assuming they
are free to do so will force price toward equilibrium. In this
equilibrium, the quantity demanded will equal the quantity
supplied.

A hypothetical numerical example may be helpful. Assume
that consumers in a market will buy zero units at $50, one at
$49, two at $48, and so on—all per month. Then the demand
is just fifty units minus the price in dollars. Since we require
as many conditions of equilibrium as there are unknown variables, we have to consider the supply schedule in order to determine what the price will be. Assume no units will be supplied at $5, two at $6, four at $7, and so on—all per month.

Then the quantity supplied is just twice the price in dollars, minus ten units.

At equilibrium, the amount supplied and demanded is equal. By finding where the supply and demand curves intersect, or by setting the two relations for quantity equal to each other and solving the algebra for the unknown price, the equilibrium price and quantity can be found. The answer for this case is that the price is $20 and the quantity is thirty units.

Obviously, buyers would be willing to buy more at a lower price, and producers would be willing to supply more than thirty units a month at a higher price. Let us suppose for the moment that the price somehow falls to $10. Then the quantity demanded would be forty units, but the quantity supplied would be only ten. Buyers will not be able to purchase all they desire, i.e., the quantity demanded at the $10 price is less than the quantity supplied. Competition among buyers will force the price up again. Conversely, if the price rose to $30, the quantity demand would be twenty units, but fifty units would be supplied. The quantity supplied, therefore, would exceed the quantity demanded, and competition among sellers would force the price down again.

Taxes

It is also possible to explain the effects of a tax given the supply and demand conditions. Continuing with our hypothetical linear demand and supply relations, let us now impose a tax on the producers. Let the supply function, or schedule, represent the costs of the producers for producing n units. Let the tax be $3. Before, no units were supplied at $5, two units at $6, and so on. With the tax, no units will be supplied at $8, two units at $9, and so on. This means that the new supply function is twice the price in dollars minus sixteen units. The demand function is, of course, still the same, fifty units minus the price in dollars, since the buyer couldn't care less about taxes on the sellers.

If your algebra is in good shape, you can now see that the equilibrium (supply equals demand) comes at a price of $22 and a demand of twenty-eight units. Otherwise, please take it on faith and observe and compare the new equilibrium price with the tax ($22) to the equilibrium price without the tax ($20). The price increase is $2, and the tax is $3. What happened?

This rather interesting outcome is what economists like to call the "incidence" of the tax. That is, who carries the burden—producers or consumers? In our simple example, a $3 tax shifts (decreases) the supply curve parallel to the old supply curve by this amount. There is no reason to believe that demand has changed. At the $20 price consumers are willing to buy thirty units. They neither know (unless they hear or read that a tax has been imposed) nor care that the producers must pay this $3 tax per unit. The whole supply curve is shifted upward and to the left because of the tax—a new supply schedule. Obviously, a new equilibrium price is determined by the intersection of the original demand and the new supply curves. Who pays the tax? In this case, both the producer and the consumer pay the tax.

The producers receive only $19 ($22 − $3), rather than $20 per unit. The consumer also shares in the burden because the price has not fallen by as much as the tax. To the consumer, these units now cost $19 plus the $3 tax, or $22 in all. Because of the nature of consumer demand (inelastic), they pay two-thirds of the tax, or $2, and the producers one-third, or $1. The tax is shifted forward onto the consumer. Had demand been very elastic or horizontal, most or all of the tax would have fallen on the producer.

In the main, a tax is shifted onto the consumer when demand is inelastic, or it is carried by the producers when supply is relatively more inelastic. Another important point is that a tax will raise prices most and reduce quantities produced less when both supply and demand are inelastic. Conversely, when they both are elastic price changes little and quantity changes much.

With the apparatus of demand and supply and its underlying forces, the economist can analyze the incidence of various taxes, subsidies, bounties, etc. A few such cases are done in Appendix C; however, we need to present a word of caution about using these tools in a static context.

Statics and Dynamics

So far in our supply and demand analysis we have found the equilibrium price by solving linear equations with constant coefficients. In other words, our analysis has been static in nature. To use an analogy, we have taken a picture in time period one and then another in time period two. This is *comparative statics*. We have not attempted to show the path by which a new equilibrium is reached or the dynamics of the system. Certainly, this is an important facet of economic analysis, especially in regard to ecostabilization policy.

For example, in order to judge the effectiveness of a set of institutions and policies in promoting ecostabilization associated with the equilibration of private costs and social costs, the economist should consider how these policies and institutions affect the ecological system while it is in the process of adjustment. To do this would require a welfare maximizing course—course in the economic system being defined as "the whole set of output, consumption and price time paths for every individual and firm in the economy." [2] This, for the economist, is an existing theoretical exercise, but requires data and dynamic modeling simply not available nor invented.

239

We have in the United States today a wealth of data on various ecodestabilizing products and a great deal of analysis could be performed to determine the effect of a given event, e.g., an increase in population, a tax, inventions, etc. And even if the analysis is mostly of a static nature, we can often obtain sufficiently correct results, results which conform to observations. Certainly, dynamic factors, or the changing paths along which various policies will direct the system, force us to do this when applying static welfare criteria to a dynamic world. These elements should be taken into account whenever possible. We, therefore, make two modest assumptions: (1) Policies that promise substantial long-run benefits are probably sound, in a prima facie sense, on a dynamic basis. (2) The system is stable, i.e., whatever the time factor (zero to infinity) adjustment will be towards equilibrium.[3] If the above two assumptions hold, then the economist can go a long way in making recommendations (his normative position) for ecostabilization. As discussed in Appendix C, the planner should try to ensure that policies are carried out with enough caution so that the quasi-static hypothesis ("stability") still holds. Given our ignorance of dynamic effects, a rational cybernetic manipulation of the economy must be based on small steps, with constant checks of feedback modifications.

APPENDIX B
EVALUATION
OF ECOLOGICAL
EXTERNALITIES

Many difficult problems come up when we try to make decisions about the economics of the ecological crisis. One of the worst problems is that of evaluating the worth of activities and processes that have no established value in the market place. Activities such as breathing fresh air or boating on a wilderness river obviously have value to us as human beings. If we want to compare these activities to marketable commodities, for purposes of making a choice in our planning efforts, we need to be able to express these activities in money terms. In the language of economic projects, we want to be able to assign quantitative benefits to these public goods that are external to the usual economic market processes.

This problem has been attacked in several ways, as mentioned in Chapter 5. First of all, in what we may call the semiquantitative approach, intuitive quantitative values have been given to such factors as, for example, the panoramic quality of the river valley and the amount of wildlife present, for the purpose of evaluating such natural environmental features as wild rivers. This approach has been used with undetermined success in the comparison of benefits from various types of water projects. A more fundamental approach is to connect the environmental externality with some type of marketable economic factor or consumer good. In following this type of fundamental economic approach, we can try to give values to explicit commodities, such as health or recreation, or we can value a mixed bag of commodities indirectly through the capitalization of such values in the form of changes in land prices.

We present here a few quite arbitrarily chosen examples of studies in which typical externalities are treated in a quantitative form. We do not present a unified overall methodology. Indeed, it is quite probable that an attempt at such a methodology would be premature, given the current status of scientific knowledge, for many difficulties remain in such research efforts, even at a quite modest level of generality. But the examples given should help at least to illustrate how progress can be made in the quantification of "qualitative" benefits. Of course, the work quoted below is a mere sample of what has been done.

Much of the research done contains nonquantitative gaps or important assumptions. Tremendous problems of analysis remain. In view of the seriousness of the ecological crisis, these problems should not be discouraging to us. Rather, we should be inspired by the results obtained to give extra attention to future research efforts.

One quantifiable benefit of environmental interest is health, another is recreation, and a third may include health, recreation, and other undefined variables, e.g., the examination of changes of land prices as they indirectly reflect human values. Health and recreation are considered here, while some discussion of the indirect methods is given in Chapter 5.

Health

Much of our practical interest in externalities, such as air pollution and water pollution, stems from concern for damages to health. A study carried out by Pyatt and Rogers in Puerto Rico [1] on benefits from the improvement of municipal water supplies is one example out of many of the explicit treatment of health benefits. The costs involved were those of building and maintaining the Commonwealth's water supply system. This water supply system was taken to provide new economic benefits by reducing diseases from water-borne organisms. Disease effects on population were broken down according to the following health categories: (1) mortality, or premature deaths; (2) morbidity, or sickness-causing absenteeism; (3) debility, or decreased working efficiency.

The effects of these disease categories were taken to modify the basic economic benefit attributable to the workers in the Puerto Rican society. This basic benefit was defined as the excess of a worker's earnings over his consumption, that is, his net earnings or net contribution to the rest of society. These earnings were calculated for the complete life span of the worker. The net present values of these earnings were calculated by discounting future earnings at some appropriate rate of interest (in the actual case, 4 percent). Since the very young and the very old had, in general, negative earnings, these earnings varied by age group, and an age group analysis was necessary for the actual numerical equations.

The basic equation for computing this value of net earnings, both for the case with and that without the water system, was based in part on earlier work.[2] The exact equation can be inspected in the original work, but the basic idea can be simply expressed. The sum of wages paid to each worker in the community was projected for each year in the future. This figure was then corrected for net earnings by subtracting the amount consumed by workers alone, as opposed to the total population. The fraction of these net earnings attributable to each age group was then calculated by using mortality tables. Finally, this "net earnings for one age group" for any year in the future was discounted. Discounting is done by dividing an amount to be received or spent n years in the future, n times by a factor of one *plus* the discount rate.

The total discounted "net earnings for one age group" is then added up for all the years that age group is still in this vale of tears. For the sake of brevity, let's call this final quantity "NEA." The presence or absence of diseases affected this NEA by decreasing the number of workers (and the number of consumers, in general) as a result of increased mortality, thereby, in general, changing any or all of the dependent variables. Losses from morbidity were counted as decreasing the payment of wages. Diseases that did not cause loss of wages were characterized as "debility" instead of morbidity and were taken as causing a decrease in production, rather than wages. In more exact terms then, debility should be taken to change these structural coefficients of the connection between labor and output, rather than the supply of labor. In the study as undertaken, effective debility was presumably directly estimated in monetary terms as it affected the value of NEA.

The actual benefit/cost ratios were then calculated by comparing the values of NEA as they would have been without the water system and as they were in actuality. The net benefits for each disease category was just the difference between these two values of NEA, summed over population age groups. To derive this difference, statistics of the department of public health in Puerto Rico were utilized on the mortality and morbidity rates of typhoid, diarrhea and enteritis, plus dysentery. It was assumed that 60 percent of observed decreases (in water-borne diseases since the inception of the water supply system) were attributable to the system. In addition, projections of the Commonwealth's population were made for future years, as well as projections of future net commonwealth income.

Since current year benefits were due, in general, to costs expended in earlier years, the benefit/cost ratio itself involved a discounted sum from the year being considered back to a base year. So the benefits were derived for each year in the future by summing the discounted differences between actual net earnings and projected net earnings as they would have been without current-year water systems expenditures, for each year back to an arbitrary base year. The costs were then the sum of the discounted costs for each of the same "string" of future years. The results for the benefit/cost ratios ranged from 0 at the base year, 1940, through values such as 0.914 for 1948, 1.2157 for 1972, and 2.313 for the year 2012.

The above study is characterized as a "preliminary appraisal," designed to develop a methodology for making decisions in constructing water systems. But the results are of general interest in that they illustrate the plausibility of developing such methodologies and of carrying out calculations based on certain types of data, such as population, aggregate income, disease mortality rates, etc., which are relatively easy to obtain.

Recreation

The social value of water recreational facilities has been the subject of numerous studies in the past few years. One particular study [3] was an investigation of the empirical nature of water use benefits. The authors, Davidson, Adams, and Seneca, emphasized the calculation of future usage, utilizing multiple regression analysis (see Chapter 8) to calculate projected demand for water sports facilities. This demand, of course, is the vital step in determining the size of the benefits

245

derived from the preservation or enhancement of water sport facilities.

First of all, the authors discussed the causes of market failure for the recreational uses of a water resource facility. Following Bator,[4] they divided the causes of market failure into ownership, technical and public goods externalities, and subsidiary "modes" of failure. Parenthetically, it is instructive to consider how the Delaware estuary, which they considered for their practical case, can be analyzed under these categories. For the Delaware estuary problem is, undoubtedly, fairly typical of other problems we will face in future water environmental calculations.

In the first place, the estuary suffers from what Bator calls an ownership externality (the "enforcement mode") in that legal rights are shared among many different users. The question of technical externality also enters in that, in view of the unique nature of each water resource area, the market structure necessarily involves a type of monopoly pricing. In addition, economic "signals" fail to control a water resource market because of its indivisible nature, while the incentive to establish free prices is absent because of the presence of increasing return to scale. The estuary also involves the failure of market existence, that is, it constitutes a public goods externality. For example, off-peak demand cannot be controlled by prices, but it is a salient feature of estuary use.

Another reason for market failure is that there is such a thing as "option demand" in that citizens would presumably like to purchase future water sports facilities, but such an option cannot be satisfied by the market. Finally, there is a "learning by doing" in such activities as boating, so that the future demand for the water recreation "product" depends on the present demand, making it inappropriate to have a user charge based on present values (since present prices would not then reflect future costs, the market mechanism would be stymied in its role as a tabulator of "dollar votes").

Returning to the main problem, that of finding the demand (and benefits) for water recreation in a given area, one can always set up in principle a cost-benefit analysis framework. To measure the cost, one can include the actual cost of providing certain facilities for swimming, fishing, and boating. The cost of maintaining the clean water can also be measured; to compare these costs with benefits, Davidson et al. found it useful to measure them as they affected the supply of dissolved oxygen (DO) in the water. The presence of dissolved oxygen is necessary for some organic life, such as fish, and its absence may indicate too much life of another kind, i.e., harmful bacteria. A certain level of dissolved oxygen present will then determine whether, for example, boating is possible. A somewhat higher level is needed for fishing, while a still higher level of dissolved oxygen will make swimming possible. One can then calculate a marginal cost curve as it depends on the amount of dissolved oxygen in the water (and the costs of the necessary sports facilities). If a marginal benefit curve can then, in actuality, be calculated, the crossing of these two curves will give the point to which the dissolved oxygen level of the estuary should be brought.

As always, it is quite difficult to measure the benefits in money terms. Davidson, Adams, and Seneca calculated, as a major step in this benefit analysis, the number of "activity days" of public use that would be expected of the estuary at various levels of dissolved oxygen. To make this prediction, the demand for activity days was calibrated by the results of a survey carried out by the University of Michigan Survey Research Center. In this survey, a number of socioeconomic and locational variables were correlated with the probability of participation in swimming, fishing, and boating. The variables chosen are shown on the next two pages: [5]

VARIABLE	DESCRIPTION
Age	discrete midpoint values of class intervals
Income	discrete midpoint values of class intervals
Sex	0 if female 1 if male
Education	1 if grade school highest attainment 2 if high school highest attainment 3 if college highest attainment
Race	0 if nonwhite 1 if white
Life cycle	0 if no children in household 1 if children in household 0 if no children over 5 years old 1 if children over 5 years old
Urbanization (belt)	0 if urban location 1 if suburban or outlying location
Occupation	0 if blue collar 1 if white collar
Region	0 if Northeast or North Central 1 if West or South
Participates in fishing	0 if nonparticipant 1 if participant
Participates in swimming	0 if nonparticipant 1 if participant
Participates in boating	0 if nonparticipant 1 if participant

VARIABLE	DESCRIPTION
Water per capita	area of sport fishing water per capita by state, 1960
Expert rating of swimming facilities in primary sampling unit	discrete values of 1 through 5
Expert rating of fishing facilities in primary sampling unit	discrete values of 1 through 5
Coastal area presence	0 inland area 1 coastline of Great Lakes present

Regression analyses on the observed Michigan data were then carried out, and as an example of the type of equation which was derived, an equation for swimming participation was derived in terms of the independent variables. The equation derived yielded an R^2 (adjusted for degrees of freedom) of 0.280, which gives a measure of the probable accuracy of the linear equation ($R^2 = 0$ denotes no correlation with reality, while $R^2 = 1$ denotes exact correlation).

The equation referred to and similar equations for fishing and boating were then used to calculate participation probabilities, using values for the independent variables characteristic of the Delaware estuary region. Together with estimates or analogous correlation equations for the amount of average participation in days and appropriate figures for the population of the region, the equation then determined the total potential demand for swimming, fishing, and boating in terms of activity days. These figures could then be compared for two cases: improvement and no improvement of the Delaware River water. For both cases, there was no significant difference in swimming days, which was about 19,800,000. This result was probably due to the availability of alternate swimming locations in the Delaware region. For boating and fishing there were significant differences. For one year, without water improvement, fishing was calculated at 7,435,000 days, while fishing in improved water was calculated at 7,-490,000 days. The corresponding figures for boating were 4,-248,000 days and 4,922,000 days.

The final benefit calculations were somewhat uncertain, due again to the difficulty in evaluating activity days in money terms. However, computing the total discounted (at 5 percent) present value of the number of days in future years, the differential between the clean and dirty river was 8,-766,000 days for boating and 881,000 days for fishing. If the days were valued at $3, an illustrative calculation showed that the marginal costs of the dissolved oxygen in the water up to the levels necessary for boating could be economically justified.

It is quite probable that such a pricing level for activity days is a serious underestimate of value. Anyway, other factors, such as perhaps a breakdown of types of swimming, such as pools versus open water swimming, should be considered by a more rigorous analysis, as noted by the authors themselves.

A somewhat different approach to the problem of recreational demands and benefits has been considered by Knetsch.[6] In this study, the point of focus is not the actual socio-economic variables that determine the consumer demand for recreation, but a calculation of the costs for various types of consumers which may, in turn, enable the economist to attach imputed prices to the recreational activity.

Knetsch assumes a hypothetical case of a recreational area which is used by people from three different cities. The costs of using the area are different for each city, attributable typically to different transportation costs. The rate of visiting the recreational area (visits/1,000 population) is taken to be a linear form in the visiting cost (in dollars), i.e., visits decreased proportionately as costs increased.

Such an actual demand/capita curve can be used to derive a theoretical relation for cities having various base costs. A particular theoretical question of interest is that of determining the total demand (number of visits) for a group of cities if the cost of visiting is changed. If the cities all have differing initial costs for visiting the recreation area, then a relationship can be found relating new visiting rates to new costs of visiting. If enough cities are represented, a more or less smooth curve can be obtained for this relationship.

This hypothetical curve is useful in finding out what precisely the benefit from the recreational facility is. Since all recreational facilities tend to be monopolistic because they are inherently individual, it is argued that the correct user benefit is the total area under such a curve, or, in other words, the amount that a perfectly discriminating monopolist would receive by charging the maximum price each user would be able to pay. In addition to these user benefits, in general there are nonuser benefits that may show up in capitalized form as a rise in land values near the recreational area. Additional nonuser benefits would be the returns to commercial establishments which depend on trade from the recreation users. To avoid double counting, only the net value added by these commercial establishments, above the amount they pay in the land costs themselves, should be counted.

Naturally, for practical use, the formula for visits per capita must be modified by other factors. Such additional factors might be income, substitution effects, and congestion. Therefore, in general, a second order attempt at a formula might be given by the functional relationship including these factors.

Such a theoretical formalism can be useful in assessing the role of possible entrance fees on the total benefits to be derived from the recreational facility. In general, under the simpler form of the visit/capita equation, the collection of a fee would reduce the number of visits and, therefore, the total benefits in real output. In practice, substitution effects involving alternate recreation choices might modify that conclusion significantly. At any rate, the specification of a methodology for demand curves can be useful for consideration of such problems as the role of possible variable costs in operating the recreational area. For example, once the marginal costs associated with such variables exceed the demand curve, visits in excess of that critical demand cost too much. Therefore, it would be appropriate to set fees to prevent such submarginal visits.

Additional difficulties come into the theory with the consideration of time constraints. Many recreational areas depend upon weekend travelers for many of their visitors. Therefore, reducing costs by the elimination of existing fees, for example, would fail to elicit more visitors from certain distant cities. Therefore, increasing fees would, correspondingly, not discourage as many visitors as might be expected from the simple formulation given above.

The difficulties of such benefit analyses are obvious. The progress made (of which these few cases are only a small sample) is indeed encouraging. The results already obtained can help open up many new lines of research in a field that will be critical for our survival in the years to come.

APPENDIX C
QUANTITATIVE ECONOMIC PLANNING AND ECOLOGICAL PROBLEMS

In this appendix we will touch on some basic problems encountered in carrying out economic planning in relation to ecological problems. Since the emphasis is on planning, and particularly public planning, the role of taxes will receive special attention. Linear programming formats are used extensively, to suggest the use of this tool in the actual planning process and to aid in the visualization of results. In many cases, the problem is either too trivial or else too complicated for practical use of programming techniques. The general problems of economic theory are, of course, nonlinear. Changes in supply costs cause price changes which modifies demand, and so on. But every nonlinear theory has a linear "tangent plane" associated with it. And we feel that some simple relations in this linearized region can be of use in examining such problems as external diseconomies, profits, and taxes, at least for small enough changes in values of economic variables for some commonly encountered market situations. So we feel that the phrasing of the problem in the form of familiar mathematical techniques supplies, at a minimum, a basis for the development of more complete theories.

The Basic Decision Problem

A useful way of looking at the problem of smog, or any other ecological problem that may be viewed as an external diseconomy, is as a conflict between the decentralized (market enterprise) economy and a real or potential centralized (socialist) system.

We can write the general activity analysis for a *centralized* economy in the form of a linear program having an objective function of the form

$$T_s = \sum_j^M b_j Q_j \tag{1}$$

where T_s is the value of the output from M production processes, and the b_j represents the unit value to society of each unit of output, and the units of output are given by Q_j. In this linear approximation, neglecting changing returns to scale and feedbacks from one product to another, the economic problem, in general, is to maximize T_s under a series of constraints on available factors of production

$$r_i \geq \sum_j^M c_{ij} Q_j, i = 1, \ldots, N \tag{2}$$

where the r_i represents the total supply of scarce factors and the c_{ij} are the structure coefficients relating the units of output to the production factors employed. Again, the relation shown is very simplified, in that the substitution of inputs to produce the same outputs is neglected. But for our modest purposes, we content ourselves with this "single technology" assumption. At any rate, the problem of the socialist or centralized planner is, then, to get the maximum value of output out of the possible physical outputs, given the restrictions on scarce capital, labor, land, and management factors.

The problem in the *decentralized* market economy is somewhat different. On the one hand, the T_s still represents the total social output of the community, but the decisions in the community are not made according to the maximization of T_s. Instead, a different objective function is maximized in actuality. This new function represents the decisions of all the entrepreneurs producing the various commodities. At a very simple level of abstraction, we can represent this function by

$$T_p = \sum_j^M a_j Q_j \tag{3}$$

where the a_j represents the profits per unit to be made on each of the j commodities. The entrepreneurs choose the Q_j values, still under the constraints of equation 2, in such a way as to maximize T_p. Naturally, in practice it may be a major problem to determine the a_j values.

This contrast between the social maximum and the private maximum is particularly severe when one considers the effect of diseconomies on the system. For example, if $j = 1$ represents automobiles, the manufacturer is interested only in his unit profit, a_1, while society as a whole is interested in the *total* value of output: the manufacturer's profit plus the value of other factors added plus the effects of diseconomies, such as the emission of harmful pollutants from exhaust pipes and crankcases. All these factors are considered in the evaluation of b_1, so that in this case we would have

$$a_1 = \text{profits} \tag{4}$$
$$b_1 = \text{profits} + (\text{value added}) \text{ costs} - \text{costs of smog}$$

Now if $j = 2$ represents another kind of automobile containing an effective (for simplicity, totally effective) smog prevention system incorporated into its design, it is obvious that if the manufacturer has to absorb the cost of the new system from his assumed (noncompetitive) profits, we might have the situation shown below, where we have assumed, for the moment, that the costs of the antismog systems do not change other costs

$$a_2 = \text{profits} - \text{cost of antismog system} \tag{5}$$
$$b_2 = \text{profits} + (\text{value added}) \text{ costs}$$

It is evident, then, that society prefers $j = 2$, while the manufacturer prefers $j = 1$ types of automobiles. The problem of the economic planner in this highly simplified case is to determine how to influence the manufacturer to produce the socially more desirable commodity. Naturally, the actual attribution of cost factors and the behavior of prices, supply and demand, etc., in general, cause severe calculational problems. A few of these complications will be considered below, in the discussion of selected specific problems.

Regulation and Prohibitive Taxation

The easiest way to achieve the planning goal is by fiat. Laws can be passed forcing the manufacturer to construct all automobiles with an antismog system. An equivalent procedure is to lay a prohibitive tax on the production of undesirable types of automobiles. In the example above, the effect of a tax would be to modify a_1 so that a_1 is less than a_2.

Regulation certainly plays an important role in modern economic life, and its use is appropriate in many cases. But it can easily be seen that regulation or prohibitive taxation may not be appropriate for the general ecological problem. A primary reason that can be cited against regulation is the difficulty of enforcement. It must be conceded that the imposition of taxes, prohibitive or otherwise, may also be subject to enforcement difficulties, but, in general, taxes should have a relatively easy time of it compared to regulation. Another objection to a strictly regulatory approach lies in the tendency to regulate without regard to alternative costs. That is, arbitrarily outlawing smog-producing vehicles could make auto transportation costs prohibitively high. Often such consequences are neglected in regulatory actions. It is, therefore, appropriate to consider the role of compensatory taxation, as opposed to regulation or prohibitive taxation.

Compensatory Taxation: Ecotaxes

The purpose of compensatory taxes is not to prohibit the production of the diseconomy, either by explicit regulation or by the imposition of taxes of prohibitive size, but to place reasonable taxes on the producers of the diseconomy and, then, to compensate those adversely affected. These taxes would not necessarily eliminate the diseconomy, but they would at least pay people back for social inconvenience and damages and often would lead to some abatement effects. As already indicated, the flexibility of the tax method would allow an experimental approach to the problem so that the economy would be saved from perhaps disastrous disturbances (some disturbance, of course, in the form of irreversible fixed investments, is unavoidable). We are not speaking here of external effects that are explicitly catastrophic in nature,

257

such as the release of substances that directly kill large numbers of people. In that case, the regulatory approach is the only possible method.

The principle, then, of a compensatory tax, or "ecotax," is clear. What is not so clear is what the amount of the tax should be, who should pay the tax, and who should receive the compensation. The economics of the situation, as well as the ethics involved, is complex, but a certain amount of discussion is certainly feasible.

Taxes placed on industries affect the proportion of product values paid to various factors and also affect profits and prices. These effects will be somewhat different for monopolistic or oligopolistic situations, as contrasted with competitive industries. We look at both these cases below. The role of taxes which are placed on consumers who produce diseconomies also requires special treatment, as sketched in the main text (see Chapter 9).

Monopolistic and Oligopolistic Producers

There exists a class of industries that is peculiar in that prices may be manipulated by the producers themselves. For this reason, profit may be in excess of a normal return to management and to capital invested. The arbitrary nature of prices may then mean that an industry that is monopolistic or oligopolistic (in this price-making sense) has a producer's surplus from the sale of its products. From the point of view of the ecological economist, this implies both advantages and disadvantages for the policy of the imposing "ecotaxes." An advantage is that there may be a ready-made producer's surplus available for compensation of those damaged by externality. A disadvantage is that the prices involved are flexible not only before taxes, but, in general, after taxes. Therefore one cannot be sure that true compensation will occur. By raising prices after the tax has been imposed, the manufacturer will, in general, succeed in passing at least part of the tax directly on to the consumer. A further pricing phenomenon may obscure the issue: prices fixed by employees rather than owners of monopolistic firms may not be truly income maximizing. The reaction of such firms to new taxes may be difficult to predict.

So we see that the features of the monopolistic or oligopolistic situation are really complicated. Still it is useful for us to consider some first steps in attacking the tax and compensation problem for this case. First of all, in order to compare the social benefits with the producer's benefit we must express the b_j in terms of the a_j. This can be written as

$$b_j = a_j + f_j + e_j \tag{6}$$

where the f_j represents the value added by other factors to the product and e_j represents the external economy produced by the commodity for activity j (in most of the cases with which we are concerned, diseconomies rather than economies are treated, so that e_j will have a negative value). If, as before, the subscript 1 denotes the production of smog-producing automobiles, while subscript 2 denotes the production of nonsmog-producing automobiles, then we can consider the imposition of a tax on smog-producing automobiles by adding a prime to the b_1, namely

$$b_1' = a_1' + t_1 + f_1 + e_1 \tag{7}$$

where a_1' is the profit after the tax is imposed, and t_1 is the tax. We can also show the smog-controlled system as

$$b_2 = a_2 + f_2 + e_2 \tag{8}$$

where, for generality, the presence of an external economy or diseconomy is still included. If there is no price change before and after the imposition of the tax, of course, we have $b_1 = b_1'$.

The question now remains, how large should the t_1 be? One answer generally offered is that the t_1 should exactly compensate the e_1—that the producer should exactly compensate the public for the externality imposed on it. This amounts to requiring that the producer pay the social marginal cost [1] of this diseconomy and can be expressed by

$$t_1 = -e_1 \qquad (9)$$

Naturally, the portion of the externality attributable to the factors f_j will then, in general, be passed on to them.

One objection to this method of taxation might be on the grounds of equity. In our linear approximation, the nonlinear effect of the externality is not shown. However, in practice, we must recognize that, for example, the smog from a million cars is more than 10 times as bad as the smog from 100 thousand cars. Historically, this has meant that producers have made investments that involve externalities in ignorance of the damage which would be caused. They, therefore, have an argument that they should not be forced to pay for externalities observable only through hindsight.

Another, and perhaps more serious argument from the social point of view is that the amount of tax may be large in relation to other costs and profits. The imposition of the large tax may then disarrange the economy greatly. Certainly, the producer himself might well be injured, but in the long run the public itself may suffer more. It might be possible to tax automobiles so that they are too costly to produce. This indeed may be one (drastic) solution to the photochemical smog problem. In the short run, however, such an overnight solution would cause severe dislocations in transportation costs and, therefore, in other industries and consumer markets (in a more general treatment, these interindustry effects would be included as additional externalities in the problem).

So, we can suggest that, in the first place, taxes should be limited to less than a quantity that severely dislocates the economy, or in very formal terms

$$t_1 \leq t_1^0 \qquad (10)$$

where the t_1^0 represents some tax maximum that does not cause excessive dislocation of the economy. The establishment of such a limit, however, would require investigation and perhaps some experimentation. Incidentally, such a limit might be regarded as a "political constraint." If a legislature compromises between, for example, the requests of lobbyists for consumer groups and lobbyists for manufacturers, then the maximum tax t_1^0 might represent the compromise tax limit. This limit is then a measurement in some sense of c-power (see Chapter 5).

For various reasons, then, we might want to begin taxation at low levels and increase them up to a certain equitable maximum. This maximum could be chosen to be the value of the social marginal product; however, in a situation of restricted competition, this maximum tax (called "the tax" below) could be greater (or less) than the amount of the diseconomy.

The ethics of the situation is complicated. Various schemes could be devised, but, as an example, we might suggest comparing the taxed case with what would happen in the regulated case. In our formalism, that means comparing the production of $j = 1$ to that of $j = 2$. If we assume the tax is not prohibitive, then we grant that a_1' is still greater than a_2, so that the commodity 1 is produced instead of 2. In that case, society is receiving a value b_1 instead of the more desirable b_2. The distribution of income to various sectors of society, however, is somewhat changed, because of the tax. It is possible to compare the value returned to the producer and that to the rest of society in both the realized case of $j = 1$ and the unrealized case of $j = 2$. If we call R_j the ratio of the producers' profit to the return to other factors in society (from just automobile production alone), then we have for $j = 1$ (and constant prices, denoted by the superscript zero):

$$R_1{}^0 = a_1' / (f_1 + t_1 + e_1) = (a_1 - t) / (f_1 + t_1 + e_1) \quad (11)$$

and for $j = 2$:

$$R_2{}^0 = a_2 / (f_2 + e_2) \quad (12)$$

A possible way of setting the tax is to require that the ratio of the producer's profits to the value returned to the rest of society be a certain fraction β of the ratio that would hold if regulation actually forced the producer to the more "social" production, $j = 2$ or

$$R_1 = \beta R_2 \quad (13)$$

If all the profits, labor and other factor costs, and external economics can be determined, we can then determine t_1 by combining equations 11, 12 and 13 for any given value of β. The producer and society can then be said to share the burden of the diseconomy according to the ratio β. Naturally, setting the numerical value of the ratio β may be a controversial political or administrative process!

The value for the t_1 of the tax can be written down immediately from the equations above, but it is perhaps better to first generalize the problem a bit. In the first place, the relation between f_1 and f_2 needs comment. f_2, in general, reflects the necessary addition of costs of diseconomy control (smog-control devices, in the example). In a single-commodity world, the f_2 could be written down unambiguously. But in the actual world, the factors added for smog control may not represent a new value added. If they are produced by idle resources, they do reflect new output, but if the resources are taken from other activities, they do not. We can express this algebraically as

$$f_2 = f_1 + \zeta \, \Delta f \tag{14}$$

where Δf is the value added for the smog control devices, while ζ takes on values between 0 and 1. If only idle resources are utilized, then $\zeta = 1$, if only nonidle, $\zeta = 0$. Other values of ζ, of course, correspond to intermediate cases. It is understood here that f_2 represents the *net* return to factors in the smog-controlled case.

Another complication is the fact that industry, in the presence of reduced or nonexistent competition, may be able to raise prices to compensate for the new tax. Therefore, in general we must write

$$a_1' = a_1 - t_1 + \Delta P_1 \tag{15}$$

where ΔP_1 is the price change from the tax as imposed. With the corrections implicit in equations 14 and 15, we can then solve for equation 13 to get the amount of the tax:

$$\tag{16}$$

$$t_1 = \frac{(a_1 + \Delta P_1)(f_2 + e_2) - \beta \, a_2 \, (f_1 + e_1)}{f_2 + e_2 + \beta \, a_2}$$

One more item of information can be utilized, namely the connection between a_1 and a_2. If, in the ideal case, there were no price change, then a_1 and a_2 would differ only by the extra amounts paid for smog control costs. If we also suppose that the producer might be able to raise prices also in the case where he actually changed over to the smog-controlled production ($j = 2$), we can write the hypothetical ideal profit in terms of the actual profit, smog control cost, and possible price change in the ideal case as

$$a_2 = a_1 - \Delta f + \Delta P_2 \qquad (17)$$

where the ΔP_2 represents the hypothetical price change. The system of equations 14 through 17 is now complete, and we can use it to determine any tax on the basis of the special welfare criterion chosen, namely that the *ratio* of return from auto production to entrepreneurs versus the rest of society should be, in the taxed and compensated case, a given fraction of the ideal smog controlled case.

This criterion contrasts with the more usual one of maximizing net benefits. The difference between choices of criteria corresponds to differing viewpoints of equity, or income distribution.

It is perhaps useful to look at the somewhat opaque expression of equation 16 in a few limiting cases. First of all, let us assume that all the prices, by accident or design, are fixed, $\Delta P_i = 0$. Then, for simplicity, let any diseconomy in the second case be negligible, so that $e_2 = 0$, and assume the welfare criterion is chosen so that $\beta = 1$. Also let us specify that e_1 is a diseconomy, so that we can write

$$e_1 = - |e_1| \qquad (18)$$

With these assumptions, we can first consider the special case $\zeta = 0$, in which the smog controlled measures would require resources from other sectors of the industry, thus causing decreased production elsewhere. In that case we have

$$t_1 = |e_1| - f_1 \cdot (|e_1| - \Delta f) / (a_1 + f_1 - \Delta f) \qquad (19)$$

where the tax is shown to be the amount of the diseconomy, minus some positive number which, for very small extra factor inputs Δf corresponds to the part of the social marginal product which is chargeable to factors other than entrepreneurs. The last factor is positive because b_2 is greater than b_1 by hypothesis (compare equations 3 and 4).

For the case in which entirely idle resources are used, it may be instructive to write the equation in a somewhat different fashion:

$$\zeta = 1: t_1 = \Delta f + |e_1| \, [1 - (f_1 + \Delta f) / (a_1 + f_1)] \qquad (20)$$

In this case the tax can be interpreted as recompensing society for the fact that the ideal case would add a net extra market economic output to the economy, in the form of the extra factor Δf. The tax would make a charge for the diseconomy e_1 also, but this charge would be reduced again by a factor that represents the social marginal product (negative) attributable to other factors in production. Note again that the last factor is positive, since it is assumed in equation 17 that there is at least a minimum profit a_2 which can be obtained in the smog-controlled case, and, therefore, $a_1 > \Delta f$.

The specific behavior of equations 19 and 20 depends on the particular values assumed, but the following general behavior may be noted. Considering the no-idle-resource case, we see that the tax is a monotonic function of $|e_1|$ and of Δf specifically. For fixed Δf, the tax increases linearly with $|e_1|$, so that larger externalities imply proportionately larger taxes. For fixed $|e_1|$, the tax increases with an increase in Δf, if the externality per unit is less than the value of the unit product. This means that for very small extra factor inputs required in the ideal case, the tax on the producer is just the part of the externality chargeable to excess profit, or producer's surplus. For larger extra factor inputs (e.g., expensive smog devices), the tax increases nonlinearly. This result may appear to be unfair in that the producer is taxed more when the solution to the externality problem would cost him more. But the equations reflect the ethical standpoint that the ideal monopolistic profit, or producer's surplus, would be the present profit reduced by the extra smog-control factor costs (this, of course, falls short of the usual ideal of economists, i.e., no excess profits). The large extra factor costs, which are not paid in fact, are taken in this view to be the *source* of the extra profits, so that the monopolist is really "cheating" society by using an inferior (e.g., smog-producing) process. One might call this point of view a tax on "adulteration of atmosphere," analogous to direct adulteration of products (like adding chalky water to milk in the bad old days of Upton Sinclair).

If the externality value is larger than the product value, then the formula gives results that imply taxes equal to or greater than the producer's surplus. Such taxes are possible, but no longer compensatory, rather they may be prohibitive—equivalent to regulation. Alternately, they may act similarly to taxes placed on a competitive industry in that marginal producers are driven out of business.

Naturally, the case of no price change, which we have treated in most detail here, represents a somewhat unrealistic situation, at least for the classical rational monopolist. Usually, prices will rise to compensate for increased marginal costs, and demand for the product $j = 1$ will decrease. Furthermore, the arguments given neglect entirely the problem of possible tax-induced equilibria between $j = 1$ and $j = 2$ production, maintained by price differentials. Of course, the simplified linear argument should give a conservative estimate of the benefits from the tax program. That is, in addition to the compensation benefits or income redistribution effects shown in the linear theory, the nonlinear effect of taxation on monopolies would act to discourage production of the externality producing goods by raising prices and so decreasing demand.

In order to get at this important effect that, in general, causes a decrease in pollution-causing production, we can abandon linearity for a moment, and hypothesize some arbitrary price-demand relationship for the production processes 1 and 2. For example, let us assume the linear demand relationship

$$Q_j = Q_{mj} \left(1 - P_j / P_{mj}\right) \tag{21}$$

where Q_{mj} is the demand at zero price for product j and P_{mj} is the price at zero quantity demand. Then, if the prices, as before, indicate the variables after the tax scheme has been introduced, we can find the change in production if we can assume that the monopolist is classically rational, that is, he wants to maximize his profit. Then, as indicated in Appendix A, since his new income after taxes, assuming other unit costs constant in the quantity interval considered, at price P_1' and quantity Q_1' is

$$I = (P_1' - t_1) Q_1' \tag{22}$$

it is maximized by setting

$$P_1' = P_1 + \tfrac{1}{2} t_1 \tag{23}$$

At this maximizing price, the new quantity produced is

$$Q_1' = Q_1 (P_1 - \tfrac{1}{2} t_1) / P_1 \tag{24}$$

and the transfer payment or compensatory "ecotax" now collected from the producer is just

267

$$t_1 \, Q_1{}' = t_1 \, Q_1 \, (P_1 - \tfrac{1}{2} t_1) \, / \, P_1 \qquad (25)$$

instead of $t_1 \, Q_1$, as in the linear case where quantity was taken as independent of price. But now we have, in addition, an actual decrease in the amount of externality produced, that is

$$|e_1| \, (Q_1 - Q_1{}') = \tfrac{1}{2} t_1 \, |e_1| \, Q_1 \, / \, P_1 \qquad (26)$$

The transfer payment plus the actual decrease in externality produced represent the value transferred to society at large from the producer as a result of the tax program.

The role of labor and other factors in the industry producing the diseconomy is treated in this formulation by requiring society to absorb that part of the burden. Certainly, it may be that, by a change of wage rates or by special employment taxes, this burden could be placed directly on labor and other factors in the affected industry. Then the tax would include the fractions of the marginal social cost applicable to these factors, and the equations above would be modified accordingly (such a case is considered in Chapter 9 of the text, to make comparison with the competitive case easier). However, in the presence of monopolistic labor unions and other factors causing stickiness in wages, such "pass-down" effects may prove to be ineffective. Since the decision power presumably lies with the entrepreneur, a general spreading of the social cost, as shown in the equations above, may be as equitable an arrangement as any in many practical cases.

We have stressed that the particular tax scheme illustrated here is based on an arbitrary ethical assumption, formalized in equation 13. But the general principles of the analysis are applicable to other, different assumptions. In practice, such ethical decisions in our society are made by pluralistic political processes.

Competitive Industries

In perfect competitive industries, the situation is somewhat different. There is ideally no producer's surplus to be utilized in paying the costs of compensatory taxes. That is, there is no surplus on an average basis, although individual competitors may generate surpluses from more efficient procedures.

At any rate, one can assume on an industry-wide basis that the effect of any tax, in general, changes the cost curves of the competitive business in such a way as to shift the supply curve of that industry (see Appendix A). Specific taxes, such as Henry George's single tax scheme, may indeed leave the supply curves unchanged, and more efficient firms may be able to absorb the extra costs from surplus revenue. But if we consider the general case, we see that the shifted supply curve will then require a higher price to be paid by the consumer for an equivalent quantity of demand, or that given the usual types of demand curves, the industry will stabilize at a lower level of production. If the taxes are distributed in such a way that consumer incomes remain approximately the same, however, the demand not supplied in the taxed industry will, under ideal circumstances, be picked up by additional demand in industries producing substitute goods. For example, it might occur that the pollution caused by the lumber industry could be reduced by taxes that discourage the production of lumber and instead stimulate increased output of cement by the substitution effect. The net result would be, then, to merely decrease the diseconomy caused by lumber (assuming cement is a "clean" product).

With this in mind, we can assume that output including substitution is constant, and we can construct an objective function for the economies and diseconomies alone, seeking to maximize the function

$$T_c = \sum_j e_j Q_j \qquad (27)$$

where the e_j are the economies or diseconomies per unit of production, and the Q_j are the quantities produced, as above. We can treat this as a linear programming problem, if we assume that the economies are constant per unit of production and that the Q_j is given by some expression linear in price, such as equation 21, where the Q_{mj} is the demand at zero price, and the P_{mj} is the maximum price, while P_j is any other feasible price. The P_j is now defined in terms of the decision variables, which are the taxes t_j, by

$$P_j = t_j + f_j \qquad (28)$$

where the other quantities are the profit and the factor costs as defined above.

The problem is not yet complete, since we must suppose that, as mentioned before, the value received in taxes must be such as not to unbalance the economy in any way so that the assumptions of pure competition (including divisibility and open entry) become radically incorrect. One way of doing this is to restrict the taxes to certain percentages of particular outputs. In that case, there is no need for the linear program, and the answer for the P_j may be written down immediately, in terms of a percentage of the price or in terms of a certain minimum quantity Q_j as determined by equations 21 and 28. Another case, in which the linear programming aspect is not entirely trivial, is that in which the taxes are taken to vary from industry to industry and are, in addition, not restricted individually, but only in toto. In that case we can write

$$\sum_j t_j \le t^0 \qquad (29)$$

270

where t^0 is the maximum amount of taxes that can be allowed. As before, this maximum amount might represent either our economic caution, as rational planners, or a reality of the political scene, the "marketplace of power," for in some sense more power will be exerted by those producers taxed against the imposition of a new tax proposed. This amount of power may be difficult to measure, but the fact of some type of proportionality must be evident. This is, of course, a question for future sociological research.

Now this maximum amount of taxation can conceivably be expressed as a fraction of the total value of the output before tax, which we can write as

$$t^0 = a \sum_j P_j^0 Q_j^0 \qquad (30)$$

where a is the chosen fraction, and $P_j^0 Q_j^0$ are the original pretax output values.

Another alternative is to set the tax at the amount of the social marginal product. In that case, the t_j can be taken as just equal to the negative of the e_j. Again, ideally, the tax in this case would be shared between the entrepreneur and the other factors in the taxed industry.

Not all industries are either monopolistic or perfectly competitive, of course. As discussed briefly in Chapter 2, various modifications of oligopoly and monopolistically competitive firms form an important sector, maybe the most important, in the actual marketplace. But we have not treated these intermediate firms explicitly here, since uncertainties in the two extreme cases are great enough as is. Certainly, a tax policy based on a faction of social marginal product is a plausible way to start, regardless of the exact price mechanisms involved in the market. And since the form of the demand curves of any real firm is difficult to determine anyway, it may be that the nature of the particular market involved in

271

any one case will have to be found out experimentally during the course of the tax program. For as "ecotaxes" are gradually raised, we can tell by the behavior of prices and the disappearance or nondisappearance of firms, whether there was a producer's surplus (excess profit) there in the first place. Then we can adjust the tax policy to: (1) get the environmental job done and (2) minimize economic dislocation in the process.

Reciprocal Externalities

Some theoretical interest has centered on the role of several independent producers operating as free agents who all produce externalities in addition to economic products. These external economies or diseconomies, in general, affect the profitability of each of the participants in a nonlinear way. A standard example is that of several factories that dispose of waste water into a lake from which they also draw fresh water supplies. The loss from diseconomies for one factory depends on the actions of each other factory. This is a topic which is appropriately treated by game theory.

For two factories, the decision problem can be represented by the following equations:

$$T_R = \sum_{ij} C_{ij} Q_i^1 Q_j^2 \tag{31}$$

where C_{ij} represents the profit to factory 1 given in units of products i and j produced by factory 1 and factory 2, and the Q_i^1 and Q_j^2 represent the quantities produced. The total quantity produced is usually restricted by the scarcity of resources, so that in shortened form, we can write production restrictions formally, in terms of factory output limits Q_m^i, as

$$\sum Q_i{}^1 \leqq Q_m{}^1 \qquad\qquad (32)$$

$$\sum Q_i{}^2 \leqq Q_m{}^2$$

Thus, if the values for factory 2 say, are given, the best production decisions for factory 1 can be determined by linear programming methods.

Unfortunately, for more than two factories, the problem becomes much more difficult. It has been shown [2] that there is a set, or "core" of solutions for this problem, only if external economies only are involved. For the case of diseconomies, such a general theorem has not yet been proved. As far as social purposes are concerned, the existence or nonexistence of this core implies that there will be some sort of rationality of overall factory operation, given the rationality of individual decisions. This fact still does not tell us how to apply taxing policies in a desirable fashion.

Some of the difficulty of this reciprocal externality problem can be avoided if cooperation between certain factories to the detriment of others can be prevented. So part of the answer may lie in the simplest answer to ecological problems, that of regulation or prohibitive taxation, since the exact answers for the shaping of these decision processes seems obscure at the present stage of research.

Factor Costs and Externalities

The diseconomies normally present in situations with disturbed ecosystems will not only change the value of social outputs in the economy, examined from the extended point of view we are considering here, but will also affect the supply curve for factors necessary for production. An obvious example is the role of air or water pollution in causing sickness. This sickness can be considered from the point of view of a negative benefit for the ill person. But here we should like to consider only the effect of increased sickness on labor hours. In terms of the general problem given in equations 1 through 3, the externality in this case causes not only a discrepancy between a_j and b_j, but also changes the value of the r_i in equation 2. We can then write the social optimum function, using the equivalent of equation 6 (with the addition of tax) and equation 28, as

$$T_s = \sum_j P_j Q_j + \sum_j e_j Q_j \qquad (33)$$

where the externality is explicitly shown. The producer optimization function is the same as before, in equation 3. The explicit form of equations 2 and 3 are now, at a high level of aggregation, and relabeling the subscript i in terms of capital K, land R, and labor L:

$$\sum_j C_{Kj} Q_j \le r_K \qquad (34)$$

$$\sum_j C_{Rj} Q_j \le r_R$$

$$\sum_j \epsilon_{Lj} Q_j + \sum_j C_{Lj} Q_j \le r_L$$

The first two equations are the usual form of equation 2. In the last equation, however, we have displayed explicitly a term ϵ_{Lj}, which gives the per unit decrease of labor supply due to the externality. That is, in the absence of the externality, r_L is the total labor supply available and C_{Lj} gives the productivity of commodity j in terms of units of labor.

The problem is now somewhat more complicated than previously. Before, the producer had to maximize equation 3 with a given labor supply, land, and capital, but he did not have to take the externality explicitly into account. The externality problem was raised in terms of the efforts of society to insure that the optimization decisions in equation 3, insofar as possible, optimize equation 1 or 33 also. Now the producer himself must reckon with the fact that his choices will be modified explicitly by the externality in that in effect he has fewer workers available for the job or, in practical terms, that labor will cost more. Naturally, this effect may show up in other categories of resources too, such as a change in the price of water for industrial purposes, due to pollution. The change would emerge in this case as an effective decrease in the supply of capital goods.

Economic Development Problems

When we are examining the structure of existing industries, it is useful for mathematical purposes to consider the linearized problem as discussed above. Such an approximation appears to be useful for the planning of tax policies in established industries. When we come to the question of economic development, it is perhaps better to recognize that nonlinearities may be of critical importance. The effect referred to here is that a very small quantity of industrial haze may be of negligible importance to anyone, while 10 times the amount of haze or smog may represent a diseconomy that appears to be 100 times worse to society. When one is planning to establish industry in a virgin area, this effect may be of great importance.

To illustrate this, we consider the case of a profit function for society in which the returns to ordinary market processes are still approximated by constant returns to scale or linearity of return to output. Let the effect of an externality now be approximated by a quadratic term:

$$T_s = \sum_j P_j Q_j + \sum_j e_j Q_j^2 / Q_j^0 \qquad (35)$$

where the Q_j^0 is some arbitrary scale factor with dimensions of commodity units in order that the other symbols retain the same units as before.

If we consider only one j value and ignore the exact form of the constraints for a moment, we can see that if e_j is negative, there will be a definite maximum of social benefit at a certain maximum value of Q_j. All of this reflects the fact that at a certain point the evils of output will begin to outweigh the benefits. For the general case of many commodities and many production constraints, the optimization may be somewhat more difficult to effect. In principle, the problem is as before; there is a conflict between the optimization goals of the producers and those of the society, but in practice it may be possible to consider this case as a simple quadratic programming problem. For example, if the attitude of the government is a key factor in the extent and form of the economic development, then all decision variables may be put in terms of governmental supplying of capital and taxing of production factors. In that case, if the government objective function is taken to be the same as that of society, analogs of equation 2 may be written in terms of possible government budget expenditures and taxes, and the quadratic programming problem for equation 35 may be solved to determine correct government policies.

Of course, the effects of externalities may be, in practice, more likely to be determined by the square of the sums of the Q_j instead of the sum of the squares. But this change causes no mathematical difficulty in the optimization problem, and the corrected objective function for the square of sums may be substituted readily for equation 35.

Stock Conservation

It has been suggested by Boulding [3] that in the society of the future the amount of social output or "throughput" may not represent the quantity to be optimized. Presumably, in future times output will be large enough so that small changes in the standard of living will not be very important. At the same time, conservation of key minerals, water and gas (i.e., air) resources will be of utmost importance. In that case, one could suppose that the goal of society would be to maximize an objective form

$$T_s = \sum_{i=1}^{N'} P_i \, \Delta r_i \tag{36}$$

where Δr_i is the change (defined as positive for an increase) in the stock resource commodity i and can be given by an equation connecting the production commodities Q_j and a generalized structure coefficient:

$$\Delta r_i = -\sum_{j=1}^{M} c'_{ij} Q_j, \, i = 1, 2, \ldots, N' \tag{37}$$

In equation 36 and 37, primes are shown on N' and c'_{ij} to emphasize that the accounting system for resources and commodities may be somewhat changed from equations 1 and 2, in that all resources necessary for the ultimate maintenance of the ecoeconomy (including presently "free goods," such as clean air) must be included in the N' and that the c'_{ij} may

277

be positive or negative, according to whether the production of commodity j decreases or increases the supply of resource i. A further possible constraint on the system might be that a certain level of production commodities must be maintained, or

$$\sum_{j=1}^{M} P_j Q_j \geqq c_0 \qquad (38)$$

where c_0 is the required commodity level for a certain standard of living in the future "resource-conserving" world. The optimization of equation 36, as defined in equation 37 and constrained by equation 38, is also restricted by the fact that production cannot exceed available resource supplies r_i in any case:

$$r_i^0 \geqq - \Delta r_i' , i = 1, \ldots , N' \qquad (39)$$

The problem still arises, in this society of the future, of how to effect the optimization of equation 36 when producers will presumably still be trying to optimize something on the order of equation 3. At least on an a priori basis the taxing and regulatory problems in such a world would seem to require a markedly different treatment. Of course, the scarcity of resources will, as usual, be reflected in the prices P_i, and some similarities in public and private profit should still be operative. We leave the difficulties of this case to future treatment.

Public Sector Expenditures

Much interest in the environmental problem has centered on expenditures made by governments themselves in building such things as sewage plants and in establishing national parks. In the present treatment we have concentrated on the role of the public sector in affecting private market decisions. The combined public-private problem is complex. The public problem itself is not quite as complex in principle, though it certainly may be in practice. The benefits from the building of sewage treatment versus water treatment plants in a river estuary, for example, require the calculation of possible benefits in terms of the monetary value of unpolluted water versus the cost of building and operating various plants. Areas of possible projects must be investigated, and the cost-benefit for each must be calculated, utilizing proper discounting procedures.

This particular problem has been treated in the literature.[4] In particular, Horowitz and Mobasheri[5] have considered explicitly the role of recreational benefits in the cost tradeoffs. The objective function they choose contains three terms which express the costs of, respectively, sewage plant treatment, water supply treatment, and lost recreational benefits. These terms can be written formally as:

$$T = \sum_{k=0}^{\infty} (a_k\, x^k + b_k\, y^k + c_k\, z^k) \qquad (40)$$

where x is the efficiency of sewage treatment plants, y is the level of pollution of water entering the water treatment plants, and z is a measure of the loss of recreational facilities, in this case measured by the deficit in dissolved oxygen (DO) in the water. The indices k represent an expansion of the costs in terms of power series in the x, y, and z. The a_k, b_k, and c_k are coefficients in this series which convert efficiency into sewage costs, pollution (measured here by the BOD, biochemical oxygen demand, of the stream water) level into water treatment costs, and finally DO deficiencies into recreational costs. Since costs, rather than benefits, are displayed, the objective function T is to be minimized. This minimization is to be effected by choosing the efficiency of the sewage plants appropriately (one variable is sufficient because water treatment quality is taken as fixed, as are the characteristics of the stream).

The actual minimization method used was a nonlinear programming scheme called the "sequential unconstrained minimization technique."

The coefficients shown in equation 40 depend on a complex variety of factors: economies of scale, flow into plants, the shape of sewage and water treatment cost curves, configuration of the river basin, etc. Also, the variables are related, as we have already asserted on intuitive grounds, by constraint equations that connect sewage upstream with BOD content downstream, while inflow and outflow of water, aeration regeneration of DO, etc. play their roles in the physical system. So we have functions of the very generalized form

$$x = x \, (\text{BOD in, BOD out}) \qquad (41)$$

where the efficiency naturally depends on the ratio of intake BOD to effluent BOD. Similarly, we have

$$y = y \, (\text{BOD}) \qquad (42)$$

where the water treatment variable depends on (or is) the BOD in the intake (which itself depends on stream factors), and

$$z = z \, (\Delta\text{DO}) \qquad (43)$$

where the cost to recreation is given in terms of the deficit ΔDO in dissolved oxygen in the stream.

Despite these complexities, the basic idea is simple: there is a trade-off in costs between sewage and water treatment

280

and inclusion of recreational costs affects this trade-off. In the study quoted, the recreational costs were estimated at between $6 and $24 per linear foot of stream. As one would expect intuitively, these costs raised the recommended expenditures for sewage treatment and lowered (proportionately) the water treatment costs. We could argue about the amount of recreational costs, but we could not argue that they do not exist. As reasonable men, we could use such mathematical programs to help us tell "how much does what" to save our environment.

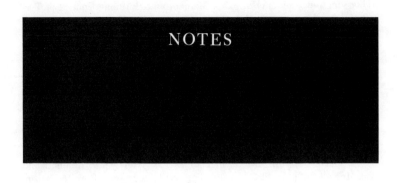

NOTES

Chapter 1

1. For the astronomical facts of life see, for example, Otto Struve, *Elementary Astronomy* (New York: Oxford University Press, 1955).

2. For an introduction to the universe, cosmogony, galaxies, etc., see the father of the steady-state (standing still) universe, Fred Hoyle, *Galaxies, Nuclei, and Quasars* (New York: Harper and Row, 1965).

3. Harrison Brown, *The Challenge of Man's Future* (New York: Viking Press, 1954).

4. See Carlo Cipolla, *The Economic History of World Population* (Baltimore: Penguin Books, 1962).

5. For evidence of what happens when man manipulates his fellow human beings, see Everett Shostrom, *Man the Manipulator* (New York: Bantam Books, 1968).

6. See Cipolla, *World Population*.

7. W. W. Rostow, *The Stages of Economic Growth* (Cambridge: Cambridge University Press, 1960).

8. Desmond Morris, *The Naked Ape* (New York: Dell Publishing Co., Inc., 1967).

9. Robert Ardrey, *African Genesis* (New York: Dell Publishing Co., Inc., 1967).

10. Robert Ardrey, *The Territorial Imperative* (New York: Dell Publishing Co., Inc., 1968).

11. For an excellent review of Ardrey's book, see Dilbert D. Thiessen, "Territory, That's What It's All About," *Contemporary Psychology*

13, no. 2 (1968); 53–54. For the reader interested in raising (if state law permits) and observing Mongolian gerbils, see G. Lindzey, D. D. Thiessen, and Ann Tucker, "Development and Hormone Control of Territorial Marking in the Male Mongolian Gerbil," *Development Psychobiology* (1968); D. D. Thiessen, H. C. Friend, and G. Lindzey, "Androgen Control of Territorial Marking in the Mongolian Gerbil (*Meriones unguiculatus*)," *Science* 160 (1968); 432–434; D. D. Thiessen, G. Lindzey, S. Blum, and Ann Tucker, "Visual Behavior of the Mongolian Gerbil," *Psychonomic Science* (1968); D. D. Thiessen, "The Roots of Territorial Marking in the Mongolian Gerbil: A Problem of Species-Common Topography," *Behavioral Research Methods and Instruments* (1968); and D. D. Thiessen, G. Lindzey, and H. C. Friend, "Spontaneous Seizures in the Mongolian Gerbil (*Meriones unguiculatus*)," *Psychonomic Science* (1968).

12. Garrett Hardin, *Nature and Man's Fate* (New York: Rinehart and Co., Inc., 1959); see also Hardin's popular biology text, Garrett Hardin, *Biology, Its Human Implications* (San Francisco: W. H. Freeman and Co., 1954); and Garrett Hardin, *Population, Evolution, and Birth Control: A Collage of Controversial Readings* (San Francisco: W. H. Freeman and Co., 1964).

13. Sigmund Freud, *Civilization and its Discontents,* ed. and trans. James Strachey (New York: W. W. Norton and Co., Inc. 1962).

14. See, for example, John A. Hobson, *The Evolution of Modern Capitalism* (New York: Scribner, 1926); Max Weber, *The Protestant Ethic and the Spirit of Capitalism* (New York: Scribner, 1952); Henri Sée, *Modern Capitalism* (New York: Adelphi, 1928). The Protestants may have gotten too much credit. For the seminal role of Catholic refugees in the seventeenth century's flowering of commerce see H. R. Trever-Roper, *The Crisis of the Seventeenth Century* (New York: Harper and Row, 1968); 5 ff.

15. Norbert Wiener, *Cybernetics* (New York: John Wiley and Sons, 1948); see also Norbert Wiener, *The Human Use of Human Beings* (New York: Doubleday and Co., Inc., 1954).

16. See Hardin, *Nature and Man's Fate,* Lindzey, Thiessen, and Tucker, "Hormone Control," Thiessen, Friend, and Lindzey, "Androgen Control," Thiessen, Lindzey, Blum, and Tucker, "Visual Behavior," Thiessen, "Territorial Marking," and Thiessen, Lindzey, and Friend, "Spontaneous Seizures."

17. For the reader who wants to obtain a working knowledge of cybernetics see W. Ross Ashby, *An Introduction to Cybernetics* (New York: John Wiley and Sons, Inc., 1963).

18. William Paddock and Paul Paddock, *Famine 1975! America's Decision: Who Will Survive?* (Boston: Little Brown and Co., 19); see also Food and Agriculture Organization of the United Nations, *Third World Food Survey* (Italy: United Nations, 1963).

19. Quoted in Robert L. Heilbroner, *The Worldly Philosophers* (New York: Simon and Schuster, 1961).

20. There are some rather important human relationships that cannot be ignored here. First off, Godwin had a best seller on his hands primarily because of a chapter in his book which argued for the abolishment of the institution of marriage. A young lady, Mary Wollstonecraft, shared his views, and they lived together in perfect harmony. However, in her sixth month of pregnancy she felt, for the sake of the child, that they should conform to the conventions of society. Of course, her father, who periodically pounded on Godwin's door, was a continuing source of encouragement. They had a daughter who was later to marry a man by the name of Shelley and was also to author the book entitled *Frankenstein*. Facts of personal history should never be ignored.

21. From the genealogical tables in the *Old Testament*, Archbishop Ussher established the date of the creation of the world as 4004 B.C. Universally accepted in England as the beginning of things, it was inserted in the King James Version of the *Bible*. (Newton also accepted the account of the early history of mankind and from a careful examination of classical authors, vague astronomical data, and so on, proceeded to built up a consistent chronology of the other great ancient kingdoms, which required, e.g., that two millennia of Egyptian history be compressed into a single generation. Newton also did other things. For a review of the life and accomplishments of Newton see L. T. More, *Isaac Newton* (New York: Scribners, 1934).

22. See Hardin, *Collage*.

Chapter 2

1. Heilbroner, *The Worldly Philosophers* (New York: Simon and Schuster, 1961).

2. Theodore W. Schultz, "Resources for Higher Education," *Journal of Political Economy* 76, no. 3 (May–June 1968): 327–347.

3. Adam Smith, *An Inquiry into the Nature and Causes of the Wealth of Nations* (New York: Random House, Inc., 1937).

4. See George J. Stigler, *The Theory of Price* (New York: Macmillan Co., 1959).

5. John Kenneth Galbraith, *The Affluent Society* (Cambridge: Houghton Mifflin Co., 1960).

6. See Harry Schwartz, *The Soviet Economy Since Stalin* (Philadelphia: Lippincott, 1965).

7. *Ibid.*, p. 142.

8. See, for example, the first essay in T. C. Koopmans, *Three Essays on the State of Economic Science* (New York: McGraw-Hill, 1957).

9. See *Newsweek*, 11 May 1970, p. 62.

10. For some of the more important articles in the field see American Economic Association, *Readings in Industrial Organization and Public Policy*, ed. R. B. Heflobower and G. W. Stocking (Homewood: Richard D. Irwin, Inc., 1958); see also National Bureau of Economic Research, *Business Concentration and Price Policy* (Princeton: Princeton University Press, 1955).

11. J. M. Clark, "Toward a Concept of Workable Competition," *American Economic Review* 30, no. 2 (June 1940): 241–256.

12. John von Neumann and Oskar Morgenstern, *The Theory of Games and Economic Behavior* (Princeton: Princeton University Press, 1947).

13. See Robert Dorfman, Paul A. Samuelson, and Robert M. Solow, *Linear Programming and Economic Analysis* (New York: McGraw-Hill Book Co., 1958).

14. Edward Chamberlin, *The Theory of Monopolistic Competition*, 7th ed. (Cambridge: Harvard University Press, 1956).

15. See Paul B. Sears, *The Living Landscape* (New York: Basic Books, Inc., 1966).

Chapter 3

1. Originally, we wanted to refer to an article on twenty-five years of smog in the *Los Angeles Times*. After one of us, however, spent a week at Lake Arrowhead, in the mountains near Los Angeles, over 5,000 feet above sea level, and suffered through two days of eye-irritating smog there, we thought the reference might be sufficiently self-explanatory to Southern Californians, and perhaps after the smog attacks on the East Coast this year, to Easterners as well.

Chapter 4

1. W. W. Leontief, *Structure of the American Economy, 1919–1929* (New York: Oxford University Press, 1951).

2. See W. W. Leontief, "Environmental Repercussions and the Economic Structure: An Input-Output Approach," *The Review Of Economics And Statistics* volume 3, no. 3 (August 1970): 262–269 where he shows how "extermatities" can be incorporated into the conventional input-output picture, and how I/O computations provide "concrete replies to some of the fundamental factual questions that should be asked and answered before a practical solution can be found to problems raised by the undesirable environmental effects of modern technology and uncontrolled economic growth."

3. J. R. Hicks, *Value and Capital*, 2nd ed. (London: Oxford University Press, 1946).

Notes

4. See Armen A. Alchian, "The Meaning of Utility Measurement," *American Economic Review* 43, no. 1 (March 1953): 26–50; Milton Friedman and L. J. Savage, "The Utility Analysis of Choices Involving Risk," *Journal of Political Economy* 56 (1948): 279–304; and Peter C. Fishburn, *Utility Theory for Decision Making* (New York: Wiley-Interscience: 1970).

5. This example (together with the basic principles of economics) is presented in Paul Samuelson, *Economics* (New York: McGraw-Hill, 1970); see also Armen A. Alchian and William R. Allen, *University Economics* (Belmont: Wadsworth Publishing Co., Inc., 1968).

6. See Milton Friedman, "The Methodology of Positive Economics," *Essays in Positive Economics* (Chicago: University of Chicago Press, 1953): 3–46.

7. See Tiber Scitovsky, *Welfare and Competition* (Chicago: Richard D. Irwin, Inc., 1951).

8. A. C. Pigou, *Economics of Welfare* (London: Macmillan and Co., 1932).

9. See *Wall Street Journal*, 7 August 1970.

10. Sidney Hook, ed., *Human Values and Economic Policy, a Symposium* (New York: New York University Press, 1967).

Chapter 5

1. For the role of fashion as the "poor man's art," which rushes to fill in cultural vacuums brought about by technological displacments, see Marshall McLuhan and Quentin Fiore, *War and Peace in the Global Village* (New York: Bantam Books, 1968).

2. See the symposium papers in Sidney Hook, ed., *Human Values and Economic Policy, a Symposium* (New York: New York University Press, 1967).

3. See the article by Richard Brandt, *ibid.*, pp. 22–40.

4. For a review of these points in the general context of welfare economics, see Tiber Scitovsky, "The State of Welfare Economics," *American Economic Review* (June 1951): 303–315.

5. See the article by Paul Samuelson in Hook, *Human Values*, p. 49.

6. See the article by Kenneth Boulding in Hook, *Human Values*, pp. 55–72.

7. Scitovsky, *Welfare Economics*.

8. See the article by Kenneth Boulding in Hook, *Human Values*, p. 68.

9. *Ibid.*, p. 71.

10. See the article by Peter Albin in Hook, *Human Values*, pp. 94–100.

11. See the article by Jon Ladd in Hook, *Human Values*, pp. 151–169.

12. *Ibid.*, pp. 163–164.

13. See the article by Milton Friedman in Hook, *Human Values,* pp. 85–93.

14. See Ronald G. Ridker, *Economic Costs of Air Pollution: Studies in Measurement* (New York: F. A. Praeger, 1967); and Sanford Rose, "The Economics of Environmental Quality," *The Environment,* eds. *Fortune* (New York: Harper and Row, 1970).

15. R^2 (or multiple determination coefficient) values for three different choices of independent variables were 0.870 and 0.939. Roughly, this should be judged on a scale between zero (false) and one (true).

16. See Jack Knetsch, "Outdoor Recreational Demands and Benefits," *Land Economics* (November 1963): 387 ff.

17. See P. Davidson, F. Adams, and J. Seneca, "The Social Value of Water Recreation Facilities Resulting from an Improvement in Water Quality: The Delaware Estuary," *Water Research,* eds. A. Kneese and Stephen C. Smith (Washington: Johns Hopkins University Press, 1966).

18. Various alternative criteria may be used. See A. Burton Weisbrod, *Economics of Public Health* (Philadelphia: University of Pennsylvania Press, 1961). Another criterion, which parallels the usual NNP per capita measures, is total output per capita. The choice involves ethical alternatives. It should be noted, however, that in practical cases in developed countries, the difference may be slight. In underdeveloped areas, if population increases faster than total output, health measures may be contraindicated on the basis of the total output per capita criterion. But the "left-over" output from workers may increase fast enough to match population growth and, therefore, that criterion may help justify additional health measures.

19. See Selma Mushkin, "Health as an Investment," *Journal of Political Economy* 70 (1962): 129 ff.

20. Edwin E. Pyatt and Peter P. Rogers, "On Estimating Benefit-Cost Ratios for Water Supply Investments," *American Journal of Public Health* (October 1962): 1, 729–1,742.

21. See Ridker, *Air Pollution,* p. 30 ff.

22. See the article by Peter Albin in Hook, *Human Values,* p. 97 ff.

23. Gene Marine, *America the Raped* (New York: Avon Books, 1969).

24. I. L. Whitman, *Uses of Small Urban River Valleys* (Baltimore: District Corps of Engineers, April 1968).

25. L. B. Leopold, "Landscape Esthetics," *Natural History Magazine* (October 1969), pp. 36–45. (Quoted in Ref. 26.)

26. Water Resources Engineers, Inc., "Fifth Progress Report: Development of Methods for Valuing Wild Rivers," Report to Office of Water Resources Research, U.S. Dept. of the Interior, Walnut Creek, Calif., January 1970.

27. See Samuelson in Hook, *Human Values,* p. 49.

28. W. Isard and C. P. Rydell, in article in *Comparative Theories of*

Social Change, ed. Hollis Peters (Ann Arbor: Foundation for Research on Social Behavior, 1966).

29. Marine, *America the Raped,* p. 35.

30. *Ibid.,* p. 32.

Chapter 6

1. Romain Gary, *The Roots of Heaven* (New York: Simon and Schuster, 1958).

Chapter 7

1. One of us spent the summer in Southeast Asia photographing and writing up environmental conditions in LDC's (less developed countries). To cite a couple of examples: The growing incidence of a new eye infection in Saigon which may or may not be correlated with increased traffic emissions; the contamination and use of Bangkok's canals as sewer ways; and the "brown" air of Tokyo as a prelude to another Donora (see Chapter 11). In LDC's their argument is: "When my child is starving I do not think about the environment."

Chapter 8

1. For a real world oligopsony situation see Walter J. Mead, *Competition and Oligopsony in the Douglas Fir Lumber Industry* (Berkeley: University of California Press, 1966).

2. For a rigorous explanation of linear programming and its uses in economics, see John von Neumann and Oskar Morgenstern, *The Theory of Games and Economic Behavior* (Princeton: Princeton University Press, 1947).

3. See "IBM Symposium on Computer Techniques Applied to Agriculture" San Jose, IBM Corporation, 1967.

4. For elementary treatments of linear, nonlinear, and dynamic programming see Charles Carr and C. W. Howe, *Quantitative Decision Procedures in Management and Economics* (New York: McGraw-Hill, 1964).

5. See Kenneth J. Arrow, "Alternative Approaches to the Theory of Choice in Risk Taking Situations," *Econometrica* 19 (October 1951): 404–437.

6. Von Neumann and Morgenstern, *Theory of Games.*

7. Straightforward treatments of regression analysis can be found in many texts. For example, see John Freund and Frank Williams, *Modern Business Statistics* (Englewood Cliffs: Prentice-Hall, 1958).

8. See the article by Kenneth Arrow in A. Kneese and Stephen C. Smith, eds., *Water Research* (Washington: Johns Hopkins University Press, 1966).

Chapter 9

1. See Fred C. Shapiro, "E. D.," *New Yorker*, 23 May 1970, pp. 93 ff.

2. R. V. Ayres and A. V. Kneese, "Production, Consumption and Externalities," *American Economic Review* 59, no. 3 (June 1969).

3. See Harry Schwartz, *The Soviet Economy Since Stalin* (Philadelphia: Lippincott, 1965).

4. See R. Lipsey and K. Lancaster, "The General Theory of Second Best," *Review of Economic Studies* 24 (November 1956): 11–32.

5. See J. Mishan, "Welfare Criteria for External Effects," *American Economic Review* 51, no. 4 (September 1961): 594–613.

6. See Allen V. Kneese, *The Economics of Regional Water Quality Management* (Washington: Johns Hopkins University Press, 1964). For recent developments, see Allen V. Kneese, *Managing Water Resources: Economics, Technology, Institutions* (Washington: Johns Hopkins University Press, 1968).

7. Economists often reason that payments are just as good as taxes or charges in that the total costs of control (clean up) and damages (diseconomies) can be minimized by judicious use of either device. But this argument makes the assumption that the interests of the firm are identical with the interests of society. Such is definitely not the case when, as often happens, the firm involved is monopolistic or oligopolistic. Then, the clean up costs could be extracted from a producer's surplus, while payments could add on to such excess profits. Things are not all rosy either in the purely competitive case. For example, to be consistent with the principle of the optimum allocation of resources, payments would have to be continued indefinitely even to the owners of a competitive firm which has given up and gone out of business! See Kneese, *Regional Water* and Kneese, *Managing Water Resources*. It follows that setting pollution charges or standards so as to equate marginal control costs and marginal damage costs may be a useful principle for public projects, but may lead to severe inequities in the actual, as opposed to the idealized "Adamistic" private sector.

8. See Chamberlin, reference 14, chapter 2.

9. See Paul Ehrlich, *The Population Bomb* (New York: Ballantine Books, 1968).

10. See the article by Kenneth Boulding in Sidney Hook, ed., *Human Values and Economic Policy, a Symposium* (New York: New York University 1967).

Chapter 10

1. See Sara Davidson, "Open Land," *Harper's* (June 1970).

Chapter 11

1. See, for example, National Commission on Community Health Services, *Changing Environmental Hazards* (Washington, D.C.: Public Affairs Press, 1967).

2. Rachel Carson, *Silent Spring* (Boston: Houghton Mifflin Co., 1962).

3. See H. L. Penman, "The Water Cycle," *Scientific American* 223, no. 3, (September 1970). The theme of this issue of *Scientific American* is devoted to presenting an understanding of the biosphere.

4. See National Commission on Community Health Services, *Changing Environmental Hazards*.

5. The number of average cholera deaths in India over the period 1946–1955 was around 80,000; however, as a result of control and immunization, the annual figure has fallen to around 14,000. In 1965, cholera spread to 23 countries where provisional figures should approximate 51,000 cases of the disease and 14,000 deaths. The World Health Organization warns that effective control measures have proved difficult, e.g., the effectiveness of vaccines has proved to be low and of short duration. We can expect a recrudescence in these times of civil commotion, mass population movements, and consequent decline of medical services. For more details see Gordon Wolstenholm and Maeve O'Connor, eds., *Symposium on Health of Mankind* (London: J & A Churchill Ltd., 1967).

6. See Edwin D. Kilbourne and Wilson G. Simillie, eds., *Human Ecology and Public Health* (London: Macmillan and Co., 1969).

7. See National Commission on Community Health Services, *Changing Environmental Hazards*.

8. *Ibid.*

9. *Ibid.*

10. See "The Mercury Mess," *Time*, 28 September 1970.

11. See Wolstenholm and O'Connor, *Symposium on Health of Mankind*.

12. See Rene Dubos, *Man, Medicine and Environment* (New York: F. A. Praeger, 1968).

13. See J. Middleton and D. Clarkson, "Motor Vehicle Pollution Control," *Traffic Quarterly* (April 1961): 306–317; also quoted in Garrett De Bell, ed., *The Environmental Handbook* (New York: Ballantine Books, 1970).

Chapter 12

1. Anthony Downs, *Inside Bureaucracy* (Boston: Little Brown and Co., 1967): 32–34.

2. For another viewpoint, see Milton Friedman, "The Role of Government in Education," *Capitalism and Freedom*, (Chicago: University of Chicago Press, 1962).

3. See Downs, *Inside Bureaucracy*, pp. 38–39.

4. Talcott Parsons, "Evolutionary Universals in Society," *American Sociological Review* 29, no. 3 (June 1964): 339–357. Most of the quotation here is taken from Max Weber; references given are in the Parsons paper.

5. See Arnold Toynbee, *A Study of History* (Oxford: Oxford University Press, 1954), Vol. 10.

6. Herbert Marcuse, *One-Dimensional Man* (Boston: Beacon Press, 1964): xv.

Appendix A

1. Alfred Marshall, *Principles of Economics*, 8th ed., (New York: Macmillan Co., 1948).

2. Melvin W. Reder, *Studies in the Theory of Welfare Economics* (New York: Columbia University Press, 1963).

3. See R. G. D. Allen, *Macro-Economic Theory* (New York: St. Martin's Press, 1968).

Appendix B

1. Edwin E. Pyatt and Peter P. Rogers, "On Estimating Benefit-Cost Ratios for Water Supply Investments," *American Journal of Public Health* (October 1962), pp. 1729–1742.

2. Burton A. Weisbrod, "The Nature and Measurement of the Economic Benefits of Improvement in Public Health," (Ph. D diss., Northwestern University, 1958).

3. P. Davidson, F. Adams, and J. Seneca, "The Social Value of Water Recreation Facilities Resulting from an Improvement in Water Quality: The Delaware Estuary," in A. Kneese and C. Stephen, eds., *Water Research*, Resources for the Future, Inc., Washington, D.C.: Johns Hopkins Press, 1966.

4. F. Bator, "The Anatomy of Market Failure," *Quarterly Journal of Economics* 72 (1 August 1958).

5. Davidson, Adams, and Seneca, *Water Resources*.

6. Knetsch, *Demands and Benefits*.

Appendix C

1. See J. E. Meade, "External Economies and Diseconomies in a Competitive Situation," *Economic Journal,* 62 (1952): 54–6 f.

2. See L. Shapley and M. Shubik, "On the Core of an Economic System with Externalities," (Santa Monica: RAND Corporation, 1969); see also the discussion in Kneese, *Water Quality Management,* where examples are given on the complexities introduced into taxation (charge) or payment systems by "nonseparable" externalities. The last concept is due to O. Davis and A. Whinston, "Externalities, Welfare, and the Theory of Games," *Journal of Political Economy* (June 1962): 241 ff.

3. See the article by Kenneth Boulding in Hook, *Human Values,* pp. 55–72.

4. See D. H. Clough and M. B. Bayer, "Optimal Waste Treatment and Pollution Abatement Benefits on a Closed River System," *Journal of the Canadian Research Society* 6, no. 3 (November 1968): 155–170.

5. A. Horowitz and F. Mobasheri, "Nonlinear Programming Applied to Regional Water Quality Management," (submitted to *Water Resources Research*).

INDEX

295